Microbiology Research Advances

nova
science publishers

www.novapublishers.com

Microbiology Research Advances

Trichoderma: Taxonomy, Biodiversity and Applications
Michael S. Mouton (Editor)
2023. ISBN: 979-8-88697-946-6 (Softcover)
2023. ISBN: 979-8-88697-984-8 (eBook)

The Rhizosphere: Structure, Ecology and Significance
Somboon Tanasupawat, PhD (Editor)
Yong-Jie Xu, PhD (Editor)
2023. ISBN: 979-8-88697-674-8 (Hardcover)
2023. ISBN: 979-8-88697-810-0 (eBook)

New Research on Mycorrhizal Fungus
Qiang-Sheng Wu (Editor)
Ying-Ning Zou (Editor)
Yue-Jun He (Editor)
Nong Zhou (Editor)
2023. ISBN: 979-8-88697-637-3 (Hardcover)
2023. ISBN: 979-8-88697-662-5 (eBook)

Endophytes: Types, Potential Uses and Mechanism(s) of Action
Pragya Tiwari, PhD (Editor)
2022. ISBN: 979-8-88697-045-6 (Hardcover)
2022. ISBN: 979-8-88697-205-4 (eBook)

Endotoxins and their Importance
Arif Pandit, PhD (Editor)
R. S. Sethi, PhD (Editor)
2022. ISBN: 978-1-68507-839-3 (Softcover)
2022. ISBN: 978-1-68507-913-0 (eBook)

More information about this series can be found at
https://novapublishers.com/product-category/series/microbiology-research-advances/

Michael S. Mouton
Editor

Trichoderma

Taxonomy, Biodiversity and Applications

nova
science publishers
www.novapublishers.com

NOTICE TO THE READER

Library of Congress Cataloging-in-Publication Data

Names: Mouton, Michael S., editor.
Title: Trichoderma : taxonomy, biodiversity and applications / Michael S. Mouton (editor)
Description: New York : Nova Science Publishers, [2023] | Series: Microbiology research advances | Includes bibliographical references and index. |
Identifiers: LCCN 2023030005 (print) | LCCN 2023030006 (ebook) | ISBN 9798886979466 (Softcover) | ISBN 9798886979848 (adobe pdf)
Subjects: LCSH: Trichoderma. | Trichoderma--Scientific applications.
Classification: LCC QK625.M7 T764 2023 (print) | LCC QK625.M7 (ebook) | DDC 579.5/677--dc23/eng/20230630
LC record available at https://lccn.loc.gov/2023030005
LC ebook record available at https://lccn.loc.gov/2023030006

Published by Nova Science Publishers, Inc. † New York

Contents

Preface

This book contains five chapters that detail the uses of trichoderma. Chapter One focuses on a comprehensive revisit of applications of Trichoderma in agriculture and other fields with a futuristic perspective. Chapter Two aims to present the main bioproducts derived from Trichoderma and the primary benefits of these microorganisms in agriculture. Chapter Three explains a study where nine isolates of Trichoderma spp. were tested. Trichoderma spp. are known for their high mycoparasitic potential as biocontrol agents against M. maydis, the cause of late wilt disease in maize. Chapter Four discusses the potentiality of Trichoderma spp. as a biocontrol agent and its mechanism in plant disease management. Lastly, Chapter Five details how Trichoderma acts as a bioinsecticide.

Chapter 1

Applications of *Trichoderma*: Past, Present and Future

S. Nabanita Kumar*
and Lalitha Pappu
Department of Microbiology and Food Science and Technology,
GITAM (Deemed to be University), Visakhapatnam, Andhra Pradesh, India

Abstract

In the current global situation, with its ever-increasing human population and the damages generated by climate change, it is imperative that effective and sustainable methods be used to ensure high and quality harvests. Biological control has many advantages over chemical pesticides, including its reduced likelihood of creating resistant strains of pests. *Trichoderma spp*. can function as a direct and indirect biocontrol agent for plant growth and disease prevention. Mycoparasitism, nutrient competition, antimicrobial compound (antibiosis) production, and lytic enzymes are all the underlying mechanisms of these fungi in benefitting the host plants. *Trichoderma* is used as a biocontrol agent against several plant pathogens. *Trichoderma* species are capable of producing specific phytohormones such as auxins, ethylene, and gibberellins. They also produce antioxidants, phytoalexins and phenols, which boost the root system's potential for branching out. The influence that *Trichoderma* has on seed germination, plant morphology, and plant physiology results in a better field stand. Fungi, such as *Trichoderma*, are gaining popularity for their capacity to biodegrade hazardous substances in an effective manner. *Trichoderma* spp. can fix nitrogen, solubilize and mineralize phosphate, potassium, and other elements, producing iron chelates and

---

* Corresponding Author's Email: nsiripur@gitam.in.

In: Trichoderma: Taxonomy, Biodiversity and Applications
Editor: Michael S. Mouton
ISBN: 979-8-88697-946-6

making them available for plant absorption. There are around 285 products registered to contain *Trichoderma* spp. for agricultural uses. MAPK signalling pathways may play a role in mycoparasitism and biocontrol. Different *Trichoderma* strains also synthesize non-ribosomal peptides (NRPs), which trigger plant defence responses. *Trichoderma* spp. is one of the beneficial rhizospheric microorganisms that exhibit a variety of features that help plants recover from environmental stress. Several *Trichoderma* species are commercially accessible as biopesticides and biofertilizers but have no place in the farmer's field. Apart from the applications in agriculture, *Trichoderma* finds use in the bioremediation of inorganic compounds using biosorption, bioaccumulation, volatilization, and Phytobial remediation. This chapter is focused on a comprehensive revisit of applications of *Trichoderma* in agriculture and other fields with a futuristic perspective.

Introduction

In the current global situation, with its ever-increasing global population and the damages generated by climate change, it is imperative that effective and sustainable methods be used to ensure high-quality harvests. Increasing food production without causing any additional strain on the environment is a pressing problem today (Germer et al. 2011). Improved agricultural methods that raise yields and gradually boost soil quality are essential for meeting this problem (Cassman et al. 2003). Large agricultural losses are sustained yearly worldwide because crops are so vulnerable to diseases brought on by plant pathogens, negatively affecting production and the product's market value (Savary et al. 2012). Pathogens are responsible for an estimated 78% loss in fruit crops, 54% loss in vegetable crops, and 32% loss in grain crops (Tudi et al. 2021). Crops vulnerable to plant diseases can suffer significant losses both in the field (before harvest) and in storage (after harvest), and plant infections are blamed for this (Mesterházy, Oláh, and Popp 2020). Pathogens can be classified into several broad categories, including viruses, bacteria, oomycetes, fungi, nematodes, and parasitic plants (Hane et al. 2020). The F.A.O., the United Nations agency in charge of food and agriculture, reports that the most crucial crops in the world are frequently hit by plant mycoses (Bruinsma 2017). It is generally accepted that fungi are the single most important factor in the widespread destruction of important crops, including rice, beans, soybeans, corn, potatoes, and wheat (Holliday 1995).

Soil-borne diseases, of which fungi are the most prevalent species, are responsible for the greatest yearly losses in both natural and production environments, killing as much as one-third of all crops (Lambers et al. 2009). So far, scientists have identified more than 19,000 unique kinds of fungi that may cause crop diseases across the world (Antonelli et al. 2020). The use of naturally occurring organisms (antagonists) to suppress pathogenic activity is an attractive strategy for managing plant diseases (Prajapati et al. 2020). Biocontrol has many advantages over chemical pesticides, including its reduced likelihood of creating resistant strains of pests, its lack of environmental impact, its compatibility with organic production, and its ability to satisfy the demands of profitable markets with regard to maximum limits of chemical residues on fruits and vegetables (Bardin et al. 2015; Kredics et al. 2003). Here we will try to understand the function of *Trichoderma* as an alternative to traditional chemical-based pesticides, insecticides, and growth promoters and how it opens up gates for sustainable agriculture. It is a significant antibiotic producer, numerous strains of this genus are 'rhizosphere competent,' meaning they can digest hydrocarbons, chlorophenols chemicals, polysaccharides, and xenobiotic pesticides used in agriculture (Lo, Nelson, and Harman 1996; Contreras-Cornejo et al. 2016). The current chapter discusses the past, present and future of *Trichoderma* with respect to agriculture, sustainability and industries.

Trichoderma and Sustainability in Agriculture

Christiaan Hendrik Persoon provided a description of the genus *Trichoderma* in the year 1794 (Rahimi Tamandegani et al. 2020). The ability of *Trichoderma* species to act as biocontrol agents against plant diseases was first identified in the early 1930s, and in the years that followed, the list of diseases that could be controlled by *Trichoderma* was significantly expanded (Howell 2003). This has resulted in the commercial production of several species of *Trichoderma* around the globe for the purpose of protecting and enhancing the growth of a variety of crops, as well as the production of species of *Trichoderma* and mixtures of species in India, Israel, New Zealand, and Sweden (Woo et al. 2014; KUMAR Sanjeev, Manibhushan, and Archana 2014; López-Bucio, Pelagio-Flores, and Herrera-Estrella 2015).

However, *Trichoderma* spp., for example, has been very successfully used as a beneficial microorganism (G. E. Harman et al. 2008b; Monte 2001; Mbarki et al. 2017). This is due to the fact that *Trichoderma* spp. Inhibit a broad spectrum of fungal plant pathogens and resources, *Trichoderma* spp. may also be evaluated as potential bio-fertilizers that have a diversity of mechanisms, are excellent competitors with the microbial community within the rhizosphere, are tolerant or resistant to in order to promote sustainable agriculture and the protection of natural (G. E. Harman et al. 2004b; Keswani et al. 2014b; Marques, Martins, and Mello 2018; Joshi, Bhatt, and Bahukhandi 2010), this would minimize the amount of chemical inputs (Pal and Gardener 2006; Heydari and Pessarakli 2010). In addition, certain species of *Trichoderma* are able to mineralize organic nutrients through the production of large quantities of extracellular enzymes, which allows them to use plant residues as a source of nutritive material (Tortella, Diez, and Durán 2005; Wallenstein and Burns 2011). The prevalence of *Trichoderma* in soil under a wide range of climatic conditions is primarily attributable to the organisms' capacity to degrade a wide range of organic substrates in soil, as well as their metabolic versatility and superior competitive saprophytic ability (Akrami and Yousefi 2015; Leimbach, Hacker, and Dobrindt 2013). Microorganisms in the rhizosphere are responsible for the production of extracellular enzymes that initiate the breakdown of high-molecular-weight polymers (Bhatti, Haq, and Bhat 2017). This process can also lead to the inhibition of plant-pathogenic fungi (Kaiser et al. 2010). These species have the potential to be outstanding candidates for supplying plants with long-term induced resistance (Poveda, Abril-Urias, and Escobar 2020; Mbarki et al. 2017). *Trichoderma* spp., on the other hand, is greatly influenced by a wide variety of environmental factors, most notably temperature and the amount of water present. In addition, the low salt tolerance of *Trichoderma* strains is one of the most significant limitations of their usage as soil inoculants. This is one of the most important factors that prevent their utilization (Kredics, Antal, and Manczinger 2000). *Trichoderma* act against several plant pathogens in variety of crops (Table 1).

Soil is necessary for plant development, and maintaining its quality is key to agriculture (Khatoon et al. 2020). Degradation of soil structure and chemistry reduces productivity (Lal 1991). Poor organic matter, nutritional deficiencies, and low microbial populations hinder plant development and yield (Mimmo et al. 2014). Desertification, fertilizer leaching, and excessive conventional farming techniques have reduced microbial diversity and nutrient levels. Conventional agricultural techniques such as monoculture, tillage, and repetitive application of inorganic fertilizers reduce macro-aggregates that serve as a microhabitat for the microbial population in the soil (Diacono and Montemurro 2011). Inorganic fertilizers can bond to soil components, generating an insoluble compound. Phosphorus, nitrogen, potassium, and micronutrients are vital for plant development (Mahato, Bhuju, and Shrestha 2018). Unavailable or inaccessible nutrients reduce output (Mehetre and Mukherjee 2015). Large portions of sprayed fertilizers seep into water bodies or volatilize in the air, causing an environmental disaster (Nadarajan and Sukumaran 2021). For agroecosystem sustainability, biological strategies for preserving and boosting soil production are essential (Altieri and Rosset 1996).

Trichoderma can fix nitrogen and solubilize and mineralize phosphate, potassium, and other elements, producing iron chelates and making them available for plant absorption (Gupta et al. 2015). It accelerates chemical and metabolic processes by changing soil pH, increasing organic matter, boosting nutrients, and modifying microbial activity (S. Lee et al. 2016). *Trichoderma* proves plant development and soil microbial structure by generating chemicals and removing harmful ones (Tripathi et al. 2013b).

Table 1. *Trichoderma* against various plant pathogens

Species of *Trichoderma*	Pathogen	Plant	Reference
T. koningiopsis	*Sclerotinia sclerotiorum*	Soybeans	(Haddad et al. 2017)
T. harzianum	*Sclerotinia sclerotiorum* *Phytophthora melonis* *Phytophthora capsici*	Lettuce Cucumber Pepper	(Budge and Whipps 1991; Ozbay and Newman 2004) (Bae et al. 2011)
T. asperellum	*Pseudomonas syringae* *Magnaporthiopsis maydis* *Phytophthora capsici*	Cucumber Maize Cacao	(Degani et al. 2021; Tondje et al. 2007; Trillas et al. 2006)
T. longibrachiatum	*Fusarium oxysporum*	Tomato	(Sallam, Eraky, and Sallam 2019)
T. asperellum	*Rhizoctonia solani*	Cucumber	(Trillas et al. 2006)
T. viride	*Sclerotium rolfsii* *Aspergillus flavus* *F. oxysporum* *Peronosclerospora sorghi*	Tomato Groundnut Banana	(Ekundayo, Ekundayo, and Osinowo 2015; Syamala D.; K Sanjeev and Eswaran 2008; Sadoma, El-Sayed, and El-Moghazy 2011)
T. atroviride	*Macrophoma theicola* *Fusarium oxysporum*	Tea, Tomato	(Anita, Ponmurugan, and Babu 2012; Rao et al. 2022)
T. parareesei	*Colletotrichum* spp.	Chilly	(Yacob 2019)
T. koningii	*Sclerotium rolfsii* *Botrytis cinerea* *Rhizoctonia solani*	Tomato Strawberry Cowpea	(Alizadeh, Sharifi-Tehrani, and Hedjaroude 2007; Latunde-Dada 1991, 1993)
T. hamatum	*Pythium* spp, *Fusarium oxysporum*	Radish, Cucumber	(G. Harman, Chet, and Baker 1980; Srinon et al. 2006)

Other researchers have shown that *Trichoderma* improves soil health and plant development under inadequate or hazardous situations (Devi et al. 2020; Ahmad and Zaib 2020; Nakkeeran et al. 2020). Like improved coastal saline soil in an experiment by creating soluble sugars, amino acids, and organic acids and enhancing potassium and calcium absorption. Further, it promoted the development of antioxidant machinery, resulting in the production of enzymes such as polyphenol oxidase that reduced salt stress (Sachdev and Singh 2020; Mohammadi and Sohrabi 2012). There are around 285 products registered to contain *Trichoderma* for agricultural uses (Woo et al. 2014; Monte 2001). These pieces of data make it abundantly evident that various species of *Trichoderma* play an important part in preserving and boosting soil productivity without affecting the latter's capacity for future output (Khan et al. 2010).

Induction of Disease Resistance in Plant

It has been reported that the presence of various *Trichoderma* species in a plant's rhizosphere improves plant defense against pathogenic organisms such as viruses, bacteria, and fungi by stimulating the initiation of various resistance mechanisms, primarily induced systemic resistance, hypersensitive response, and systemic acquired resistance (Pieterse et al. 2014; G. E. Harman et al. 2004b; Mendes, Garbeva, and Raaijmakers 2013). Based on multiple observations, an inference may be made in favor of several groups of metabolites, emphasizing their importance as elicitors or resistance inducers in *Trichoderma*-plant interactions (Sood et al. 2020; Aflatoxins 2015; Sivasithamparamb et al. 2008). These metabolites contain proteins with enzymatic activity, such as xylanases and chitinases, which results in the enzymatic breakdown of fungal or plant cells (Druzhinina et al. 2011; Aflatoxins 2015).

Plants have an immune system that is able to recognize microbe-associated molecular patterns (M.A.M.P.s) (Seiboth et al. 2011; Sharfman et al. 2014). It is possible that the capacity of *Trichoderma* spp. Hyphae to produce M.A.M.P.s for molecular recognition contribute to the signal cascade that is causes used by signalling molecules found inside the plant, such as salicylic acid, jasmonic acid, and ethylene (Chakraborty, Chakraborty, and Sunar 2020; Hermosa et al. 2013). The colonization of roots by *Trichoderma* species produces a clear systemic resistance through a salicylic acid signalling cascade. Additionally, salicylic acid, jasmonic acid, and ethylene signalling pathways combine in the systemic resistance that is triggered by cell-free culture filtrates of *Trichoderma* (Salas-Marina et al. 2011). In both greenhouse and field conditions, *Trichoderma* spp. plays a vital role in preventing plant diseases and slowing the growth of disease-causing pathogens (Nagaraju et al. 2012; Akrami et al. 2012). Given that *Trichoderma* generates a wide variety of elicitors, which interact with plant receptors to cause the detection of *Trichoderma* and the induction of resistance in plants, it seems clear that *Trichoderma* is responsible for both of these processes (Hermosa et al. 2013). In addition, many strains of *Trichoderma* are capable of producing tiny secondary metabolites (Frisvad et al. 2018), which have been demonstrated to activate pathogenesis-related protein (P.R. protein) (Kauffmann et al. 1987) and to decrease the systemic elimination of the disease (Saravanakumar et al. 2016).

Promoting Plant Growth and Stress Tolerance

The yield of crops all around the world was negatively impacted by abiotic factors such as salinity, drought, the buildup of heavy metals, and severe temperatures (Gull, Lone, and Wani 2019; Suzuki et al. 2014). It has been discovered that applying *Trichoderma* species to wheat, mustard, tuberose, corn, sugarcane, tomato, and maize, among other plants, results in a higher yield (Singh et al. 2018b), this tolerance can be induced against abiotic challenges (Swain and Mukherjee 2020). Plants that have been infested by *Trichoderma* generate specific phytohormones such as auxins, ethylene, and gibberellins (Jaroszuk-Ściseł et al. 2019; Spaepen 2015). They also produce antioxidants, phytoalexins and phenols, all of which give tolerance to abiotic stressors and boost the root system's potential for branching out (Shoresh, Harman, and Mastouri 2010). *Trichoderma* is capable of producing cytokinin-like molecules (such as zeatin) and gibberellin-related molecules (such as GA3 or GA4)(Tucci et al. 2011), both of which have the potential to contribute to biological improvements in crop fertility (Guzmán-Guzmán et al. 2019). Additionally, a higher photosynthetic rate was reported in rice that had been infected with *Trichoderma* in a similar manner (G. Harman et al. 2021; Doni et al. 2014), the presence of *Trichoderma* in maize plants resulted in higher growth (Björkman, Blanchard, and Harman 1998), enhanced root biomass output (López-Bucio, Pelagio-Flores, and Herrera-Estrella 2015), and increased root hair formation (Degani and Dor 2021). A rice plant that has been treated with *Trichoderma* has improved its ability to absorb nutrients

(Cuevas 2006), rice plants that had been treated with *T. harzianum* had a dramatically higher ability to resist drought and water deficiency conditions (Shukla et al. 2012), which contributed to better nutrient absorption and plant development (Doni et al. 2014). A study that was conducted on tomato seeds that included *T. harzianum* showed that *Trichoderma* has the ability to speed up the seed germination process, alleviate water, osmotic, salinity, chilling and heat stressors, and induce physiological protection against cellular damage. Additionally, it has been observed to enhance the foliar area, facilitate secondary root formation, and change root architecture (Newman, Brown, and Ozbay 2002).

The Influence that *Trichoderma* has on seed germination, plant morphology, and plant physiology results in a better field stand, and it also speeds up the growth of plants in both their vegetative and reproductive stages (Singh et al. 2018b; N. Kumar and Khurana 2021). It results in an increase in the plant's total number of branches, spikes, flowers, and fruits. In many instances, the average weight of a single fruit is also far more than it was previously (Viera et al. 2019).

Insecticidal Activity

Trichoderma demonstrated the potential Insecticidal in many entomological studies (Ghosh and Pal 2016). It is generally done it by means of indirect and direct involvement. Through direct involvement, it parasitizes the insect and kills it eventually (Verma et al. 2007). There are many studies which include control of the Silverleaf whitefly (*Bemisia tabaci*) (Jafarbeigi et al. 2020), the tropical bed bug *(Cimex hemipterus*) (Zahran et al. 2017), the coconut palm rhinoceros beetle (*Oryctes rhinoceros*) (Nasution et al. 2018), the bean weevil *(Acanthoscelides obtectus*) (Rodríguez-González, Casquero, et al. 2018) and the vineyard borer (*Xylotrechus arvicola*) (Rodríguez-González, Carro-Huerga, et al. 2018), In addition, owing to the foliar application of fungal spores, *T. longibrachiatum* is able to increase the crop yield of eggplants by 56% in the field. This is due to the fact that it can kill up to 50% of the lepidopteran eggplant/brinjal borer *(Leucinodes orbonalis*) (Ghosh and Pal 2016). The indirect approach includes the activation of systemic response by inducing salicylic acid and jasmonic acid, which leads to an active plant difference mechanism against the insect (Martínez-Medina et al. 2017).

Composting

Trichoderma is used successfully to turn Agri waste and industrial waste into compost (Sala et al. 2021). Only a small amount of the cellulose, hemicellulose and lignin produced as byproducts in agriculture or forestry is used, the rest being considered waste (Soccol et al. 2019). Waste lignocellulose materials from agricultural, agro-industrial processing, forestry, and municipal organic waste help fulfil the requirements for inexpensive animal feeds, fuel, fertilizer and various chemicals (Datta, Hossain, and Roy 2019; Dar et al. 2021).

Industrial Uses

The ability of *Trichoderma* to create lytic enzymes is put to use in the brewing and winemaking industries, as well as in the production of animal feed (Blaszczyk et al. 2014). *Trichoderma* spp. is capable of using a diverse collection of chemicals, including carbon and nitrogen sources (Rani and Dhania 2014), while concurrently secreting a number of enzymes to break down plant polymers into simple sugars for the purposes of energy production and development (Haran, Schickler, and Chet 1996). As a result of the high cost of the chemical inducers for these enzymes (Kath et al. 2017), it is necessary to identify some low-cost organic inducers that may be derived from agricultural waste in order to increase the amount of *Trichoderma* species that are produced in large quantities (Dhillon, Kaur, and Brar 2013). It is vital to do research on the enzymes that are generated by *Trichoderma* in order to discover more efficient and cost-effective enzymes that will be beneficial in various aspects of the hydrolytic process of biomass degradation (Almeida et al. 2021).

Cellulases, hemicelluloses, and pectinases are three types of enzymes that are generated by *Trichoderma* (Olsson et al. 2003). These enzymes are utilized in the partial hydrolysis of plant cell walls in feeds, which increases the digestibility of the feed and the feed's nutritional value (Y.M. GALANTE, DE CONTI, and MONTEVERDI 1998b). It is also possible to employ enzymes derived from *Trichoderma* as food additives (Kunamneni et al. 2014), for instance, to increase the raw material maceration that takes place during the manufacturing of fruit and vegetable juices (H. P. Sharma, Patel, and Sugandha 2017). Additionally, the enzymes are utilized to increase the flavour of wine as well as the fermentation, filtration, and overall quality of the beer (Bhat 2000).

Cellulases that are generated by *Trichoderma* are used in the textile industry to soften and condition the fabrics as well as make high-quality washing powders (Y. Galante, De Conti, and Monteverdi 1998a). Cellulases are also used in the production of washing powders (Ito et al. 1989). The pulp and paper industry also makes use of enzymes derived from *Trichoderma* in order to change fibre characteristics and reduce lignin concentrations (Wong and Saddler 1992). In return, mutants produced from *T. harzianum* may be used in the formulation of toothpaste in order to aid in the reduction of plaque development (Schuster and Schmoll 2010).

T. reesei Simmons is responsible for the manufacture of cellulases and hemicellulases (Peterson and Nevalainen 2012), both of which are utilized in the generation of bioethanol from agricultural byproducts (Gray, Zhao, and Emptage 2006). These enzymes are responsible for the breakdown of substrates into simple sugars, which are then fermented by yeasts of the species *Saccharomyces cerevisiae* (Huang et al. 2023). The stimulation of *Trichoderma* hydrolytic enzymes, including pectinase, cellulase, chitinase, and glucanase activity, as well as the supra-extraction of grape, was investigated for the fermentation process for the clarifying of the juice and for the characteristics of the wine (S. Sharma et al. 2019).

Bioremediation

Chemical reduction, electrochemical treatment, ion exchange, precipitation, and evaporation recovery are the primary conventional Physicochemical procedures for removing contaminants from many environments (Crini and Badot 2010). However, these procedures have a number of severe drawbacks, including a high creation of additional harmful byproducts in addition to high energy and chemical consumption, a high cost of recovery, insufficient removal, and partial removal of the substance. In addition to completely destroying the organic chemicals, bioremediation can give a low-cost option in comparison to other remediation processes, which stabilize or get rid of the contamination (Yaashikaa et al. 2021; Varsha, Kumar, and Rathi 2022; Mousazadeh et al. 2021). Bioremediation can do both of these things successfully. *Trichoderma* spp. are exceptionally resilient against a wide variety of toxicants, including heavy metals, organometallic compounds, tannery effluents, and hazardous chemicals such as cyanide (C.N.) (Alothman et al. 2020; Mohsenzadeh and Shahrokhi 2014; V. Kumar and Dwivedi 2021; Daccò et al. 2020). Because of this, they belong to an important species of fungi that should be investigated further as a potential genetic resource for use in the bioremediation of harmful contaminants. Bioremediation of inorganic compounds is mainly done using biosorption, bioaccumulation, volatilization, and Phytobial remediation (Roy et al. 2015). Organic compounds, which are classified as a hazardous organic chemicals like Polycyclic aromatic hydrocarbons, naphthalene, phenanthrene, and benzo[a]pyrene, a crude oil spill can also be eliminated using *Trichoderma* using in vitro systems (Zafra and Cortés-Espinosa 2015). By providing these microorganisms with additional nutrients, carbon sources, or electron donors, the method of bioremediation is able to speed up the natural process by which microbes break down pollutants in their natural environment (Bollag, Mertz, and Otjen 1994). In a perfect world, bioremediation would lead to the total mineralization of pollutants into water and carbon dioxide with no intermediates being produced, but in practice, this outcome is extremely rare (Khalid et al. 2021).

Fungi, such as *Trichoderma,* are gaining popularity as a result of their capacity to biodegrade hazardous substances in an effective manner (Puyam 2016). Because of an external enzyme system that catalyzes processes that can break down hazardous aromatic chemicals, the fungi are efficient (Karigar and Rao 2011). Because they are able to degrade a wide variety of compounds, including pesticide residues in the soil, such as chlordane, lindane, and D.D.T., they are valuable for the cleanup of areas that have been polluted by pesticides (Tripathi et al. 2013a).

Characteristics of Pathogenicity of *Trichoderma*

Since the species that make up the genus *Trichoderma* are so versatile and useful, research into this group of ascomycetous fungi has received a great deal of attention (Mukherjee et al. 2013). Through the coordinated activation of systemic immune responses, *Trichoderma* spp. can function as a direct and indirect biocontrol agent, allowing for the more rapid and robust induction of plant basal resistance mechanisms in response to later triggering stimuli (Morán-Diez et al. 2021). The fungus of the genus *Trichoderma* are often found in decaying wood and other organic plant matter, and they are asexual (the teleomorphic forms are called Hypocrea) (J. M. Steyaert et al. 2013). The fungus of the genus *Trichoderma* are filamentous; they are mostly soil-dwelling and saprophytic (McAllister et al. 1994); they are avirulent, and they are opportunistic plant symbionts (G. E. Harman et al. 2004a). Mycoparasitism, nutrient competition, antimicrobial compound (antibiosis) production, and lytic enzymes are all examples of direct biocontrol mechanisms used by *Trichoderma* spp. to combat pathogens, while indirect mechanisms like inducing systemic resistance, boosting plant growth, and improving rhizosphere competence are all examples of how these fungi work to benefit host plants (Elad 2000; G. E. Harman et al. 2008a).

Mycoparasitism, antibiosis, induced disease resistance, and competitive exclusion all play a role in *Trichoderma*-mediated disease suppression (Sharon, Chet, and Spiegel 2011). When *Trichoderma* spp. Biocontrol fungi are applied to seeds or roots (Baker 1989). They quickly colonize the spermosphere (the zone around the seeds) and the rhizosphere (the zone around the roots), helping to exclude invading diseases (Raupach and Kloepper 1998). Direct antibiosis occurs when a *Trichoderma* fungus's secondary metabolites or secreted enzymes kill or prevent the development of a pathogen (Daniel and Rodrigues Filho 2007; Vinale et al. 2008). The metabolites may have a role in competitiveness, mycoparasitism, and induced disease resistance (Reino et al. 2008).

Mode of Action

The mode of action of *Trichoderma is* achieved by mycoparasitism, antibiosis, competition, and induction of disease resistance in the plant (Howell 2003).

Mycoparasitism

Mycoparasitic fungi are those that create parasitic relationships with host fungi, either living hyphae or dormant structures like sclerotia (Adams and Ayers 1979). The parasitized fungus is commonly referred to as the host or prey. *Trichoderma* spp. may exert its Mycoparasitic effects through a variety of antagonistic mechanisms, including antibiosis, competition and the production of cell wall-degrading enzymes. In later phases of the parasitic process, partial breakdown of the host cell wall is commonly observed (J. Steyaert et al. 2003).

Mycoparasitism is utilized for biological pest management, and some *Trichoderma* species are utilized as biological control agents for plant pathogens (Brimner and Boland 2003). It kills diseases and plant-associated fungi in many environments (Verma et al. 2007). A genome study revealed more complicated cell wall-degrading enzymes than expected in *Trichoderma* spp. then the line between biotroph and necrotroph is not fully crisp, but the *Trichoderma*-host fungal interaction develops aggressively from intracellular growth toward the breakdown of the host, even while an unbroken host cell wall exists (Mukherjee, Mendoza-Mendoza, et al. 2022). These mycoparasitic abilities involve chemotropism, lysis of the pathogen's cell wall (the key to mycoparasitism), hyphal penetration by appressorial formation, production of cell wall-degrading enzymes and peptaibols, mediated by heterotrimeric G-proteins and M.A.P. kinases, and parasitizing pathogen's cell wall contents - (1,6)-glucanases, chitinases, and proteases degrade pathogen cell walls during mycoparasitism, several members from each of these groups have been demonstrated to be implicated in mycoparasitism (Mukherjee, Mendoza-Mendoza, et al. 2022).

Trichoderma has a mycoparasitic impact on fungal infections by detecting and attracting prey by chemotaxis, adhering to the host, and attacking physically by branching and coiling around the hyphae of the host (Tyśkiewicz et al. 2022). Furthermore, similar to pathogen appressoria, *Trichoderma* is capable of developing appressoria-like penetrating structures (N. Lee, D'Souza, and Kronstad 2003). In the last phase of the mycoparasitic relationship, the host is killed as a result of the chemical assault and cell wall breakdown caused by hydrolytic enzymes and antifungal chemicals generated by *Trichoderma* (Rousseau et al. 1996).

Antibiosis

There has been significant discussion in the past about the biological activity of some of the secondary metabolites that are produced by *Trichoderma*. The chemicals that have antibiotic action are the primary focus of attention since it is more likely that these compounds will be involved in the success of the strain that produces them in its role as a biological control agent (Keswani et al. 2014a; Reino et al. 2008). It also seems that the synergistic impact that they have with the enzymes that they create is crucial (Tchameni et al. 2020), even while it is abundantly clear that these species have a prodigious capacity for the creation of secondary metabolites (Sachdev and Singh 2020), relatively little is known about the mechanisms that influence the production of specific metabolites (Li et al. 2018). The assumption that has been drawn from this review is that strains that produce various metabolites are actually distinct. *Trichoderma* sp. produces secondary metabolites that include antifungal chemicals from a number of chemical groups (Romano et al. 2018). They were divided into three categories:

1. Volatile antibiotics such as 6-pentyl-a-pyrone (6PP) and the majority of isocyanide derivates (Scarselletti and Faull 1994).
2. Water-soluble compounds such as peptidic acid or koningic acid (Bansal et al. 2021).
3. Peptaibols, which are linear oligopeptides of 12-22 amino acids rich in a- amino isobutyric acid, N-acetylated at the N-terminus (Vinale et al. 2012).

Trichoderma spp. metabolites (e.g., volatiles, extracellular enzymes, and antibiotics) were thought to be the likely factors implicated in antibiosis (Verma et al. 2007). *Trichoderma* sp. has also been shown to be beneficial in cases of a broad host range and in reducing the survival of pathogenic fungi (Ghildiyal and Pandey 2008), most important of these compounds are peptaibols and which are linear amphipathic polypeptide metabolites (Chutrakul et al. 2008). The biological features of these antibiotic compounds included lipid membrane disruption, antimicrobial activity and plant resistance induction (Górniak, Bartoszewski, and Króliczewski 2019).

Competition

Competition for limited resources leads to biological control of fungal phytopathogens, which is the primary cause of mortality for microorganisms (Heydari and Pessarakli 2010). For instance, iron absorption is necessary for the survivability of the vast majority of filamentous fungi (Hamilos, Samonis, and Kontoyiannis 2012). When they are iron deficient, most filamentous fungi produce low-molecular-weight ferric-iron specific chelators that are referred to as siderophores in order to mobilize environmental iron (Haas 2003). In most cases, effective antagonists are able to resist the fungistatic effect of the soil, which is caused by the presence of metabolites generated by other species, including plants, and can also persist in settings of very intense competition (Adnan et al. 2019; Benítez et al. 2004). Because *Trichoderma* strains are naturally resistant to many toxic compounds, including herbicides, fungicides, and pesticides such as D.D.T. and phenolic compounds, and can recover very rapidly after the addition of sublethal doses of some of these compounds, *Trichoderma* strains are able to grow rapidly when inoculated in soil (P. K. Sharma and Gothalwal 2017; Gajera et al. 2013). *Trichoderma* strains are also naturally resistant to phenolic compounds (Pascale et al. 2017). The capacity of *Trichoderma* spp. to get nutrients from the metabolism of carbohydrates such as cellulose, chitin, glucan, and glucose, which are frequently present in the mycelial environment, is required for optimal nutrient utilization (Sood et al. 2020). The function of the glucose transport system is unknown; however, it is possible that it plays a critical role in *Trichoderma* competition (Contreras-Cornejo et al. 2016). Root exudates and the rhizosphere are particularly rich in nutrients such as carbohydrates, amino acids, organic acids, vitamins, Fe while *Trichoderma* and pathogenic fungi compete for Carbon (Jain, Chakraborty, and Das 2020). *Trichoderma* is also responsible for the production of extremely powerful siderophores, which chelate iron and inhibit the development of other fungi (Altomare et al. 1999). Because of this, the composition of the soil has an impact on how successful *Trichoderma* is as a biocontrol agent against Pythium in relation to the availability of iron (Adnan et al. 2019). *Trichoderma* is effective in preventing the spread of *Fusarium oxysporum* because it acts as a competitor for rhizosphere colonization (Carvalho et al. 2014).

Genetics of Pathogenicity

Gene expression profiles of plant and fungal partners reveal how *Trichoderma* and host respond at the transcriptome (Villalobos-Escobedo et al. 2020), proteome (V. Sharma et al. 2017). Early gene expression alterations matter, molecular dialogue's earliest stages (Mendoza-Mendoza et al. 2018). Target genes occupy central network roles Six of 18 genes selected this way tiny cysteine-rich proteins, including Sm2 Systemic resistance effector (Mukherjee, Horwitz, et al. 2022). M.A.P.K. cascades in *Trichoderma* spp. Include M.A.P.K.K.K., M.A.P.K.K., and M.A.P.K., and M.A.P.K. signalling pathways may play a role in mycoparasitism and biocontrol (Zhang and Zhuang 2022; Zeilinger and Omann 2007). The essential parts of *Trichoderma*'s chemical arsenal for killing infections are the production and release of cell wall degrading enzymes (C.W.D.E.s) and antibiotics (Singh et al. 2018a). *Trichoderma* produces glucan and chitin synthases to repair self-cell wall damage caused by pathogens during *Trichoderma*-pathogen contact (Silva et al. 2019). Different *Trichoderma* strains also synthesize non-ribosomal peptides (N.R.P.s), which trigger plant defence responses and have antibacterial effects against many types of fungus. N.R.P.s are produced by multimodular mega-enzymes termed non-ribosomal peptide synthetases (N.R.P.S.s) outside the ribosome and are sometimes followed by secondary modifications (Mukherjee et al. 2011; Bills et al. 2014). Peptabiotics and epidithiodioxopiperazines are examples of N.R.P.s from *Trichoderma* (Contreras-Cornejo et al. 2020).

The seven transmembranes G protein-coupled receptor *Gpr1* is involved in sensing the fungal prey in Nearby, except this, there are many genes Vel1 of *Trichoderma virens* is involved in hydrophobin expression for adherence to the host and mycoparasitism (Druzhinina et al. 2011; Sarma et al. 2014). Ligand binding with such receptors leads to downstream signalling events via the activation of G-protein cascades (Omann and Zeilinger 2010).

Conclusion

Plants are sessile creatures that are continually subjected to a variety of biotic and abiotic stressors in the field, posing a threat to their life and production. The presence of two or more stressors at the same time might be damaging to plant development. For example, the presence of phytopathogens and nutrient deficiencies in the soil at the same time might have a negative impact on plant development and production, perhaps leading to crop failure. To deal with such difficult circumstances and to boost agricultural output, chemical-based procedures are commonly used, which are not sustainable, and an alternate strategy that not only minimizes the impact of stress but also preserves ecological integrity is required. Plants interact with a variety of helpful microbes in natural environments and communicate extensively with them. This helpful connection protects plants from environmental challenges and favourably controls plant development. *Trichoderma* is one of the beneficial rhizospheric microorganisms that exhibit a variety of features that help plants recover from environmental stress. Several *Trichoderma* species are commercially accessible as biopesticides and biofertilizers; however, they have no place in the farmer's field. This is owing to a lack of consistency and efficacy in field circumstances and a lack of information about molecular processes. Plants undergo physiological and biochemical changes and molecular reprogramming during plant-*Trichoderma* contact to respond to environmental stimuli and improve plant fitness. As a result, investigations of plant-*Trichoderma* interaction at the biochemical, physiological, and molecular levels can reveal the mechanism influencing plant activity under adverse environmental conditions. Secondary metabolite chemical and pharmacological investigations of several *Trichoderma* species have revealed that these fungi create a variety of chemicals with potential use in medicine, biotechnology, and agriculture. These chemical changes are generally associated with the fungal lifestyle. *Trichoderma*-plant interactions show that the fungus may influence plant development, trigger defensive responses, and solubilize inaccessible soil nutrients. The primary challenge is to develop and use in *Trichoderma* research the most recent, comprehensive, cutting-edge, and at the same time affordable, quick, and efficient methods of detecting and testing antagonists, combining multiple modes of action and causing a cascade of reactions. This is the most difficult aspect of the research.

References

Adams, PB, and WA Ayers. 1979. "Ecology of Sclerotinia species." *Phytopathology* 69 (8): 896-899.

Adnan, Muhammad, Waqar Islam, Asad Shabbir, Khalid Ali Khan, Hamed A Ghramh, Zhiqun Huang, Han YH Chen, and Guo-dong Lu. 2019. "Plant defense against fungal pathogens by antagonistic fungi with *Trichoderma* in focus." *Microbial pathogenesis* 129: 7-18.

Aflatoxins. 2015. World Health Organization; Geneva, Switzerland: World Health Organization.

Ahmad, Imtiaz, and Sania Zaib. 2020. "Mighty microbes: plant growth promoting microbes in soil health and sustainable agriculture." In *Soil Health*, 243-264. Springer.

Akrami, Mohammad, Hossein Karbalaei Khiavi, Haji Shikhlinski, and Hossein Khoshvaghtei. 2012. "Bio controlling two pathogens of chickpea *Fusarium solani* and *Fusarium oxysporum* by different combinations of *Trichoderma harzianum*, *Trichoderma asperellum* and *Trichoderma virens* under field condition." *International Journal of Agricultural Science Research* 1 (3): 41-45.

Akrami, Mohammad, and Zohreh Yousefi. 2015. "Biological control of *Fusarium wilt* of tomato (*Solanum lycopersicum*) by *Trichoderma* spp. as antagonist fungi." Biological Forum.

Alizadeh, HR, A Sharifi-Tehrani, and Gh A Hedjaroude. 2007. "Evaluation of the effects of chemical versus biological control on Botrytis cinerea agent of gray mould disease of strawberry." *Communications in Agricultural Applied Biological Sciences* 72 (4): 795-800.

Almeida, Déborah Aires, Maria Augusta Crivelente Horta, Jaire Alves Ferreira Filho, Natália Faraj Murad, and Anete Pereira de Souza. 2021. "The synergistic actions of hydrolytic genes reveal the mechanism of *Trichoderma harzianum* for cellulose degradation." *Journal of Biotechnology* 334: 1-10.

Alothman, Zeid A, Ali H Bahkali, Abdallah M Elgorban, Mohammed S Al-Otaibi, Ayman A Ghfar, Sami A Gabr, Saikh M Wabaidur, Mohamed A Habila, and Ahmed Yacine Badjah Hadj Ahmed. 2020. "Bioremediation of explosive TNT by *Trichoderma viride*." *Molecules* 25 (6): 1393.

Altieri, Miguel A, and Peter Rosset. 1996. "Agroecology and the conversion of large-scale conventional systems to sustainable management." *International Journal of environmental studies* 50 (3-4): 165-185.

Altomare, C, WA Norvell, Thomas Björkman, and GE91438 Harman. 1999. "Solubilization of phosphates and micronutrients by the plant-growth-promoting and biocontrol fungus *Trichoderma harzianum* Rifai 1295-22." *Applied and environmental microbiology* 65 (7): 2926-2933.

Anita, S, P Ponmurugan, and R Ganesh Babu. 2012. "Significance of secondary metabolites and enzymes secreted by *Trichoderma atroviride* isolates for the biological control of Phomopsis canker disease." *African Journal of Biotechnology* 11 (45): 10350-10357.

Antonelli, Alexandre, RJ Smith, C Fry, Monique SJ Simmonds, Paul J Kersey, HW Pritchard, MS Abbo, C Acedo, J Adams, and AM Ainsworth. 2020. "*State of the World's Plants and Fungi*." Royal Botanic Gardens (Kew); Sfumato Foundation.

Bae, Hanhong, Daniel P Roberts, Hyoun-Sub Lim, Mary D Strem, Soo-Chul Park, Choong-Min Ryu, Rachel L Melnick, and Bryan A Bailey. 2011. "Endophytic *Trichoderma* isolates from tropical environments delay disease onset and induce resistance against *Phytophthora capsici* in hot pepper using multiple mechanisms." *Molecular Plant-Microbe Interactions* 24 (3): 336-351.

Baker, Ralph. 1989. "Improved *Trichoderma* spp. for promoting crop productivity." *Trends in Biotechnology* 7 (2): 34-38.

Bansal, Ravindra, Shikha Pachauri, Deepa Gururajaiah, Pramod D Sherkhane, Zareen Khan, Sumit Gupta, Kaushik Banerjee, Ashish Kumar, and Prasun K Mukherjee. 2021. "Dual role of a dedicated GAPDH in the biosynthesis of volatile and non-volatile metabolites-novel insights into the regulation of secondary metabolism in *Trichoderma* virens." *Microbiological Research* 253: 126862.

Bardin, Marc, Sakhr Ajouz, Morgane Comby, Miguel Lopez-Ferber, Benoît Graillot, Myriam Siegwart, and Philippe C Nicot. 2015. "Is the efficacy of biological control against plant diseases likely to be more durable than that of chemical pesticides?" *Frontiers in Plant Science* 6: 566.

Benítez, Tahía, Ana M Rincón, M Carmen Limón, and Antonio C Codon. 2004. "Biocontrol mechanisms of *Trichoderma* strains." *International microbiology* 7 (4): 249-260.

Bhat, MK. 2000. "Cellulases and related enzymes in biotechnology." *Biotechnology advances* 18 (5): 355-383.

Bhatti, A. A., S. Haq, and R. A. Bhat. 2017. "Actinomycetes benefaction role in soil and plant health." *Microb Pathog* 111: 458-467. https://doi.org/10.1016/j.micpath.2017.09.036. https://www.ncbi.nlm.nih.gov/pubmed/28923606.

Bills, Gerald, Yan Li, Li Chen, Qun Yue, Xue-Mei Niu, and Zhiqiang An. 2014. "New insights into the echinocandins and other fungal non-ribosomal peptides and peptaibiotics." *Natural product reports* 31 (10): 1348-1375.

Björkman, Thomas, Lisa M Blanchard, and Gary E Harman. 1998. "Growth enhancement of shrunken-2 (sh2) sweet corn by *Trichoderma harzianum* 1295-22: effect of environmental stress." *Journal of the American Society for Horticultural Science* 123 (1): 35-40.

Blaszczyk, LMSKS, Marek Siwulski, Krzysztof Sobieralski, Jolanta Lisiecka, and Malgorzata Jedryczka. 2014. "*Trichoderma* spp.–application and prospects for use in organic farming and industry." *Journal of plant protection research* 54 (4).

Bollag, Jean-Marc, Tawna Mertz, and Lewis Otjen. 1994. "*Role of microorganisms in soil bioremediation*." ACS Publications.

Brimner, Theresa A, and Greg J Boland. 2003. "A review of the non-target effects of fungi used to biologically control plant diseases." *Agriculture, ecosystems & environment* 100 (1): 3-16.

Bruinsma, Jelle. 2017. *World agriculture: towards 2015/2030: an FAO perspective*. Routledge.

Budge, SP, and JaM Whipps. 1991. "Glasshouse trials of *Coniothyrium minitans* and *Trichoderma* species for the biological control of *Sclerotinia sclerotiorum* in celery and lettuce." *Plant Pathology* 40 (1): 59-66.

Carvalho, Daniel DC, Murillo Lobo Junior, Irene Martins, Peter W Inglis, and Sueli Mello. 2014. "Biological control of *Fusarium oxysporum* f. sp. phaseoli by *Trichoderma harzianum* and its use for common bean seed treatment." *Tropical Plant Pathology* 39: 384-391.

Cassman, Kenneth G, Achim Dobermann, Daniel T Walters, and Haishun Yang. 2003. "Meeting cereal demand while protecting natural resources and improving environmental quality." *Annual Review of Environment Resources* 28 (1): 315-358.

Chakraborty, BN, U Chakraborty, and K Sunar. 2020. "Induced immunity developed by *Trichoderma* species in plants." In *Trichoderma*, 125-147. Springer.

Chutrakul, Chanikul, Marcos Alcocer, Kevin Bailey, and John F Peberdy. 2008. "The production and characterisation of trichotoxin peptaibols, by *Trichoderma asperellum*." *Chemistry & biodiversity* 5 (9): 1694-1706.

Contreras-Cornejo, Hexon Angel, Lourdes Macías-Rodríguez, Ek del-Val, and John Larsen. 2020. "Interactions of *Trichoderma* with plants, insects, and plant pathogen microorganisms: chemical and molecular bases." *Co-evolution of secondary metabolites*: 263-290.

Contreras-Cornejo, Hexon Angel, Lourdes Macías-Rodríguez, EK Del-Val, and John Larsen. 2016. "Ecological functions of *Trichoderma* spp. and their secondary metabolites in the rhizosphere: interactions with plants." *FEMS microbiology ecology* 92 (4): fiw036.

Crini, Grégorio, and Pierre-Marie Badot. 2010. *Sorption processes and pollution: conventional and non-conventional sorbents for pollutant removal from wastewaters*. Presses Univ. Franche-Comté.

Cuevas, Virginia C 2006. "Soil inoculation with *Trichoderma pseudokoningii* Rifai enhances yield of rice." *Philippine Journal of Science* 135 (1): 31.

Daccò, Chiara, Lidia Nicola, Marta Elisabetta Eleonora Temporiti, Barbara Mannucci, Federica Corana, Giovanna Carpani, and Solveig Tosi. 2020. "*Trichoderma*: evaluation of its degrading abilities for the bioremediation of hydrocarbon complex mixtures." *Applied Sciences* 10 (9): 3152.

Daniel, Juliana F de S, and Edson Rodrigues Filho. 2007. "Peptaibols of *Trichoderma*." *Natural product reports* 24 (5): 1128-1141.

Dar, RA, M Parmar, EA Dar, RK Sani, and UG Phutela. 2021. "Biomethanation of agricultural residues: Potential, limitations and possible solutions." *Renewable and Sustainable Energy Reviews* 135: 110217.

Datta, Arup, Aslam Hossain, and Sanjay Roy. 2019. "An overview on biofuels and their advantages and disadvantages." *Asian Journal of Chemistry*.

Degani, Ofir, and Shlomit Dor. 2021. "*Trichoderma* biological control to protect sensitive maize hybrids against late wilt disease in the field." *Journal of Fungi* 7 (4): 315.

Degani, Ofir, Soliman Khatib, Paz Becher, Asaf Gordani, and Raviv Harris. 2021. "*Trichoderma asperellum* secreted 6-pentyl-α-pyrone to control magnaporthiopsis maydis, the Maize late wilt disease agent." *Biology* 10 (9): 897.

Devi, Rubee, Tanvir Kaur, Divjot Kour, Kusam Lata Rana, Ashok Yadav, and Ajar Nath Yadav. 2020. "Beneficial fungal communities from different habitats and their roles in plant growth promotion and soil health." *Microbial Biosystems* 5 (1): 21-47.

Dhillon, Gurpreet Singh, Surinder Kaur, and Satinder Kaur Brar. 2013. "Perspective of apple processing wastes as low-cost substrates for bioproduction of high value products: A review." *Renewable and sustainable energy reviews* 27: 789-805.

Diacono, Mariangela, and Francesco Montemurro. 2011. "Long-term effects of organic amendments on soil fertility." In *Sustainable agriculture volume 2*, 761-786. Springer.

Doni, Febri, Anizan Isahak, Che Radziah Che Mohd Zain, and Wan Mohtar Wan Yusoff. 2014. "Physiological and growth response of rice plants (*Oryza sativa* L.) to *Trichoderma* spp. inoculants." *Amb Express* 4 (1): 1-7.

Druzhinina, Irina S, Verena Seidl-Seiboth, Alfredo Herrera-Estrella, Benjamin A Horwitz, Charles M Kenerley, Enrique Monte, Prasun K Mukherjee, Susanne Zeilinger, Igor V Grigoriev, and Christian P Kubicek. 2011. "*Trichoderma*: the genomics of opportunistic success." *Nature reviews microbiology* 9 (10): 749-759.

Ekundayo, EA, FO Ekundayo, and IA Osinowo. 2015. "Antifungal activities of *Trichoderma* viride and two fungicides in controlling diseases caused by *Sclerotium rolfsii* on tomato plants." *Advances in Applied Science Research* 6 (3): 12-19.

Elad, Y. 2000. "Biological control of foliar pathogens by means of *Trichoderma harzianum* and potential modes of action." *Crop protection* 19 (8-10): 709-714.

Frisvad, Jens C, Lars LH Møller, Thomas O Larsen, Ravi Kumar, José Arnau, and Biotechnology. 2018. "Safety of the fungal workhorses of industrial biotechnology: update on the mycotoxin and secondary metabolite potential of *Aspergillus niger*, *Aspergillus oryzae*, and *Trichoderma reesei*." *Applied Microbiology* 102 (22): 9481-9515.

Gajera, Harsukh, Rinkal Domadiya, Sunil Patel, Mansukh Kapopara, and Balubhai Golakiya. 2013. "Molecular mechanism of *Trichoderma* as bio-control agents against phytopathogen system–a review." *Curr. Res. Microbiol. Biotechnol* 1 (4): 133-142.

Galante, YM, A De Conti, and R Monteverdi. 1998a. "Application of *Trichoderma* enzymes in the textile industry." *Trichoderma & Gliocladium* 2: 311-325.

Galante, Yves M, Alberto De Conti, and Riccardo Monteverdi. 1998b. "Application of *Trichoderma* enzymes." *Trichoderma and Gliocladium, Volume 2: Enzymes, Biological Control and commercial applications* 2: 327.

Germer, Jörn, Joachim Sauerborn, Folkard Asch, Jan de Boer, Jürgen Schreiber, Gerd Weber, and Joachim Müller. 2011. "Skyfarming an ecological innovation to enhance global food security." *Journal für Verbraucherschutz und Lebensmittelsicherheit* 6 (2): 237-251.

Ghildiyal, A, and A Pandey. 2008. "Isolation of cold tolerant antifungal strains of *Trichoderma* sp. from glacial sites of Indian Himalayan region." *Research Journal of Microbiology* 3 (8): 559-564.

Ghosh, Swapan Kr, and Sujoy Pal. 2016. "Entomopathogenic potential of *Trichoderma longibrachiatum* and its comparative evaluation with malathion against the insect pest Leucinodes orbonalis." *Environmental monitoring and assessment* 188 (1): 1-7.

Górniak, Ireneusz, Rafał Bartoszewski, and Jarosław Króliczewski. 2019. "Comprehensive review of antimicrobial activities of plant flavonoids." *Phytochemistry Reviews* 18 (1): 241-272.

Gray, Kevin A, Lishan Zhao, and Mark Emptage. 2006. "Bioethanol." *Current opinion in chemical biology* 10 (2): 141-146.

Gull, Audil, Ajaz Ahmad Lone, and Noor Ul Islam Wani. 2019. "Biotic and abiotic stresses in plants." *Abiotic biotic stress in plants*: 1-19.

Gupta, Govind, Shailendra Singh Parihar, Narendra Kumar Ahirwar, Sunil Kumar Snehi, and Vinod Singh. 2015. "Plant growth promoting rhizobacteria (PGPR): current and future prospects for development of sustainable agriculture." *J Microb Biochem Technol* 7 (2): 096-102.

Guzmán-Guzmán, Paulina, María Daniela Porras-Troncoso, Vianey Olmedo-Monfil, and Alfredo Herrera-Estrella. 2019. "*Trichoderma* species: versatile plant symbionts." *Phytopathology* 109 (1): 6-16.

Haas, H. 2003. "Molecular genetics of fungal siderophore biosynthesis and uptake: the role of siderophores in iron uptake and storage." *Applied microbiology and biotechnology* 62 (4): 316-330.

Haddad, Patrícia Elias, Luis Garrigós Leite, Cleusa Maria Mantovanello Lucon, and Ricardo Harakava. 2017. "Selection of *Trichoderma* spp. strains for the control of *Sclerotinia sclerotiorum* in soybean." *Pesquisa Agropecuária Brasileira* 52: 1140-1148.

Hamilos, Georgios, George Samonis, and Dimitrios P Kontoyiannis. 2012. "Recent advances in the use of Drosophila melanogaster as a model to study immunopathogenesis of medically important filamentous fungi." *International Journal of Microbiology* 2012.

Hane, James K, Jonathan Paxman, Darcy AB Jones, Richard P Oliver, and Pierre De Wit. 2020. "'CATAStrophy', a genome-informed trophic classification of filamentous plant pathogens–how many different types of filamentous plant pathogens are there?" *Frontiers in Microbiology* 10: 3088.

Haran, S, H Schickler, and I Chet. 1996. "Molecular mechanisms of lytic enzymes involved in the biocontrol activity of *Trichoderma harzianum*." *Microbiology* 142 (9): 2321-2331.

Harman, Gary E, Thomas Björkman, Kristen Ondik, and Michal Shoresh. 2008a. "Changing paradigms on the mode of action and uses of *Trichoderma* spp. for biocontrol." *Outlooks on pest management* 19 (1): 24.

---. 2008b. "Changing paradigms on the mode of action and uses of *Trichoderma* spp. for biocontrol." *J Outlooks on pest management* 19 (1): 24.

Harman, Gary E, Charles R Howell, Ada Viterbo, Ilan Chet, and Matteo Lorito. 2004a. "*Trichoderma* species—opportunistic, avirulent plant symbionts." *Nature reviews microbiology* 2 (1): 43-56.

---. 2004b. "*Trichoderma* species—opportunistic, avirulent plant symbionts." *Nature reviews microbiology* 2 (1): 43-56.

Harman, GE, I Chet, and R Baker. 1980. "*Trichoderma* hamatum effects on seed and seedling disease induced in radish and pea by *Pythium* spp. or *Rhizoctonia solani*." *Phytopathology* 70 (12): 1167-1172.

Harman, GE, Febri Doni, Ram B Khadka, and Norman Uphoff. 2021. "Endophytic strains of *Trichoderma* increase plants' photosynthetic capability." *Journal of applied microbiology* 130 (2): 529-546.

Hermosa, Rosa, M Belén Rubio, Rosa E Cardoza, Carlos Nicolás, Enrique Monte, and Santiago Gutiérrez. 2013. "The contribution of *Trichoderma* to balancing the costs of plant growth and defense." *Int. Microbiol* 16 (2): 69-80.

Heydari, Asghar, and Mohammad Pessarakli. 2010. "A review on biological control of fungal plant pathogens using microbial antagonists." *Journal of biological sciences* 10 (4): 273-290.

Holliday, Paul. 1995. *Fungus diseases of tropical crops*. Courier Corporation.

Howell, CR 2003. "Mechanisms employed by *Trichoderma* species in the biological control of plant diseases: the history and evolution of current concepts." *Plant disease* 87 (1): 4-10.

Huang, Rong, Fangfang Zhang, Hong Zhou, Hongfei Yu, Lei Shen, Jiao Jiang, Yi Qin, Yanlin Liu, and Yuyang Song. 2023. "Characterization of *Trichoderma reesei* endoglucanase displayed on the Saccharomyces cerevisiae cell surface and its effect on wine flavor in combination with β-glucosidase." *Process Biochemistry* 124: 140-149.

Ito, Susumu, Shitsuw Shikata, Katsuya Ozaki, Shuji Kawai, Kikuhiko Okamoto, Shigeo Inoue, Akira Takei, Yu-ichi Ohta, and Tomokazu Satoh. 1989. "Alkaline cellulase for laundry detergents: production by *Bacillus* sp. KSM-635 and enzymatic properties." *Agricultural and biological chemistry* 53 (5): 1275-1281.

Jafarbeigi, F, MA Samih, H Alaei, and H Shirani. 2020. "Induced tomato resistance against *Bemisia tabaci* triggered by salicylic acid, β-Aminobutyric Acid, and *Trichoderma*." *Neotropical Entomology* 49 (3): 456-467.

Jain, Akansha, Joydeep Chakraborty, and Sampa Das. 2020. "Underlying mechanism of plant–microbe crosstalk in shaping microbial ecology of the rhizosphere." *Acta physiologiae plantarum* 42 (1): 1-13.

Jaroszuk-Ściseł, Jolanta, Renata Tyśkiewicz, Artur Nowak, Ewa Ozimek, Małgorzata Majewska, Agnieszka Hanaka, Katarzyna Tyśkiewicz, Anna Pawlik, and Grzegorz Janusz. 2019. "Phytohormones (auxin, gibberellin) and ACC deaminase in vitro synthesized by the mycoparasitic *Trichoderma* DEMTkZ3A0 strain and changes in the level of auxin and plant resistance markers in wheat seedlings inoculated with this strain conidia." *International Journal of Molecular Sciences* 20 (19): 4923.

Joshi, BB, RP Bhatt, and D Bahukhandi. 2010. "Antagonistic and plant growth activity of *Trichoderma* isolates of Western Himalayas." *Journal of Environmental Biology* 31 (6): 921.

Kaiser, Christina, Marianne Koranda, Barbara Kitzler, Lucia Fuchslueger, Jörg Schnecker, Peter Schweiger, Frank Rasche, Sophie Zechmeister-Boltenstern, Angela Sessitsch, and Andreas Richter. 2010. "Belowground carbon allocation by trees drives seasonal patterns of extracellular enzyme activities by altering microbial community composition in a beech forest soil." *New Phytologist* 187 (3): 843-858.

Karigar, Chandrakant S, and Shwetha S Rao. 2011. "Role of microbial enzymes in the bioremediation of pollutants: a review." *Enzyme research* 2011.

Kath, Juliana, Cláudia R Dias-Arieira, Júlio César Antunes Ferreira, Juliana Aparecida Homiak, Camila Rocco da Silva, and Carine Rezende Cardoso. 2017. "Control of *Pratylenchus brachyurus* in soybean with *Trichoderma* spp. and resistance inducers." *Journal of Phytopathology* 165 (11-12): 791-799.

Kauffmann, Serge, Michel Legrand, Pierrette Geoffroy, and Bernard Fritig. 1987. "Biological function of 'pathogenesis-related' proteins: four PR proteins of tobacco have 1, 3-β-glucanase activity." *The EMBO journal* 6 (11): 3209-3212.

Keswani, Chetan, Sandhya Mishra, Birinchi Kumar Sarma, Surya Pratap Singh, and Harikesh Bahadur Singh. 2014a. "Unraveling the efficient applications of secondary metabolites of various *Trichoderma* spp." *Applied microbiology and biotechnology* 98 (2): 533-544.

Keswani, Chetan, Sandhya Mishra, Birinchi Kumar Sarma, Surya Pratap Singh, and Harikesh Bahadur Singh. 2014b. "Unraveling the efficient applications of secondary metabolites of various *Trichoderma* spp." *Applied microbiology biotechnology advances* 98 (2): 533-544.

Khalid, Farah Eryssa, Zheng Syuen Lim, Suriana Sabri, Claudio Gomez-Fuentes, Azham Zulkharnain, and Siti Aqlima Ahmad. 2021. "Bioremediation of diesel contaminated marine water by bacteria: A review and bibliometric analysis." *Journal of Marine Science and Engineering* 9 (2): 155.

Khan, Mohammad Saghir, Almas Zaidi, Munees Ahemad, Mohammad Oves, Pervaze Ahmad Wani, and Soil Science. 2010. "Plant growth promotion by phosphate solubilizing fungi–current perspective." *Archives of Agronomy* 56 (1): 73-98.

Khatoon, Zobia, Suiliang Huang, Mazhar Rafique, Ali Fakhar, Muhammad Aqeel Kamran, and Gustavo Santoyo. 2020. "Unlocking the potential of plant growth-promoting rhizobacteria on soil health and the sustainability of agricultural systems." *Journal of environmental management* 273: 111118.

Kredics, László, Zsuzsanna Antal, and László Manczinger. 2000. "Influence of water potential on growth, enzyme secretion and in vitro enzyme activities of *Trichoderma harzianum* at different temperatures." *Current Microbiology* 40 (5): 310-314.

Kredics, László, Zsuzsanna Antal, László Manczinger, András Szekeres, Ferenc Kevei, and Erzsébet Nagy. 2003. "Influence of environmental parameters on *Trichoderma* strains with biocontrol potential." *Food Technology Biotechnology advances* 41 (1): 37-42.

Kumar, Narendra, and SM Paul Khurana. 2021. "*Trichoderma*-plant-pathogen interactions for benefit of agriculture and environment." In *Biocontrol Agents and Secondary Metabolites*, 41-63. Elsevier.

Kumar, Vinay, and SK Dwivedi. 2021. "Bioremediation mechanism and potential of copper by actively growing fungus *Trichoderma lixii* CR700 isolated from electroplating wastewater." *Journal of Environmental Management* 277: 111370.

Kunamneni, Adinarayana, Francisco J Plou, Miguel Alcalde, and Antonio Ballesteros. 2014. "*Trichoderma* enzymes for food industries." In *Biotechnology and biology of Trichoderma*, 339-344. Elsevier.

Lal, Rattan. 1991. "Soil structure and sustainability." *Journal of sustainable agriculture* 1 (4): 67-92.

Lambers, Hans, Christophe Mougel, Benoît Jaillard, and Philippe Hinsinger. 2009. "Plant-microbe-soil interactions in the rhizosphere: an evolutionary perspective." *Plant soil* 321 (1): 83-115.

Latunde-Dada, AO 1991. "The use of *Trichoderma koningii* in the control of web blight disease caused by *Rhizoctonia solani* in the foliage of cowpea (*Vigna unguiculata*)." *Journal of Phytopathology* 133 (3): 247-254.

---. 1993. "Biological control of southern blight disease of tomato caused by *Sclerotium rolfsii* with simplified mycelial formulations of *Trichoderma koningii*." *Plant Pathology* 42 (4): 522-529.

Lee, Nancy, Cletus A D'Souza, and James W Kronstad. 2003. "Of smuts, blasts, mildews, and blights: cAMP signaling in phytopathogenic fungi." *Annual review of phytopathology* 41: 399.

Lee, Samantha, Melanie Yap, Gregory Behringer, Richard Hung, and Joan W Bennett. 2016. "Volatile organic compounds emitted by *Trichoderma* species mediate plant growth." *Fungal biology biotechnology advances* 3 (1): 1-14.

Leimbach, Andreas, Jörg Hacker, and Ulrich Dobrindt. 2013. "E. coli as an all-rounder: the thin line between commensalism and pathogenicity." *Between pathogenicity commensalism*: 3-32.

Li, Ningxiao, Alsayed Alfiky, Wenzhao Wang, Md Islam, Khoshnood Nourollahi, Xingzhong Liu, and Seogchan Kang. 2018. "Volatile compound-mediated recognition and inhibition between *Trichoderma* biocontrol agents and *Fusarium oxysporum*." *Frontiers in Microbiology* 9: 2614.

Lo, CT, EB Nelson, and GE Harman. 1996. "Biological control of turfgrass diseases with a rhizosphere competent strain of *Trichoderma harzianum*." *Plant disease*.

López-Bucio, José, Ramón Pelagio-Flores, and Alfredo Herrera-Estrella. 2015. "*Trichoderma* as biostimulant: exploiting the multilevel properties of a plant beneficial fungus." *Scientia horticulturae* 196: 109-123.

Mahato, Sanjay, Susmita Bhuju, and Jiban Shrestha. 2018. "Effect of *Trichoderma* viride as biofertilizer on growth and yield of wheat." *Malays. J. Sustain. Agric* 2 (2): 1-5.

Marques, Eder, Irene Martins, and Sueli Correa Marques de Mello. 2018. "Antifungal potential of crude extracts of *Trichoderma* spp." *Biota Neotropica* 18.

Martínez-Medina, Ainhoa, Ivan Fernandez, Gerrit B Lok, María J Pozo, Corné MJ Pieterse, and Saskia CM Van Wees. 2017. "Shifting from priming of salicylic acid-to jasmonic acid-regulated defences by *Trichoderma* protects tomato against the root knot nematode *Meloidogyne incognita*." *New phytologist* 213 (3): 1363-1377.

Mbarki, Sonia, Artemi Cerdà, Marian Brestic, Rai Mahendra, Chedly Abdelly, and Jose Antonio Pascual. 2017. "Vineyard compost supplemented with *Trichoderma harzianum* T78 improve saline soil quality." *Land Degradation Development* 28 (3): 1028-1037.

McAllister, C áB, I Garcia-Romera, A Godeas, and JA Ocampo. 1994. "Interactions between *Trichoderma koningii*, *Fusarium solani* and *Glomus mosseae*: effects on plant growth, arbuscular mycorrhizas and the saprophyte inoculants." *Soil Biology Biochemistry* 26 (10): 1363-1367.

Mehetre, Sayaji T, and Prasun K Mukherjee. 2015. "*Trichoderma* improves nutrient use efficiency in crop plants." *Nutrient use efficiency: from basics to advances*: 173-180.

Mendes, Rodrigo, Paolina Garbeva, and Jos M Raaijmakers. 2013. "The rhizosphere microbiome: significance of plant beneficial, plant pathogenic, and human pathogenic microorganisms." *FEMS microbiology reviews* 37 (5): 634-663.

Mendoza-Mendoza, Artemio, Rinat Zaid, Robert Lawry, Rosa Hermosa, Enrique Monte, Benjamin A Horwitz, and Prasun K Mukherjee. 2018. "Molecular dialogues between *Trichoderma* and roots: role of the fungal secretome." *Fungal Biology Reviews* 32 (2): 62-85.

Mesterházy, Ákos, Judit Oláh, and József Popp. 2020. "Losses in the grain supply chain: Causes and solutions." *Sustainability* 12 (6): 2342.

Mimmo, T, D Del Buono, R Terzano, N Tomasi, G Vigani, C Crecchio, R Pinton, G Zocchi, and S Cesco. 2014. "Rhizospheric organic compounds in the soil–microorganism–plant system: their role in iron availability." *European Journal of Soil Science* 65 (5): 629-642.

Mohammadi, Khosro, and Yousef Sohrabi. 2012. "Bacterial biofertilizers for sustainable crop production: a review." *J ARPN J Agric Biol Sci* 7 (5): 307-316.

Mohsenzadeh, Fariba, and Farzad Shahrokhi. 2014. "Biological removing of Cadmium from contaminated media by fungal biomass of *Trichoderma* species." *Journal of Environmental Health Science and Engineering* 12 (1): 1-7.

Monte, Enrique 2001. "Understanding *Trichoderma*: between biotechnology and microbial ecology." *International Microbiology* 4 (1): 1-4.

Morán-Diez, María E, Angel Emilio Martinez de Alba, M Belén Rubio, Rosa Hermosa, and Enrique Monte. 2021. "*Trichoderma* and the plant heritable priming responses." *Journal of Fungi* 7 (4): 318.

Mousazadeh, Milad, Elnaz Karamati Niaragh, Muhammad Usman, Saif Ullah Khan, Miguel Angel Sandoval, Zakaria Al-Qodah, Zaied Bin Khalid, Vishakha Gilhotra, and Mohammad Mahdi Emamjomeh. 2021. "A critical review of state-of-the-art electrocoagulation technique applied to COD-rich industrial wastewaters." *Environmental Science and Pollution Research* 28 (32): 43143-43172.

Mukherjee, Prasun K, Benjamin A Horwitz, Uma Shankar Singh, Mala Mukherjee, and Monika Schmoll. 2013. *Trichoderma: biology and applications*. CABI.

Mukherjee, Prasun K, Benjamin A Horwitz, Franceso Vinale, Pierre Hohmann, Lea Atanasova, and Artemio Mendoza-Mendoza. 2022. "Molecular Intricacies of *Trichoderma*-Plant-Pathogen Interactions." *Frontiers in Fungal Biology*.

Mukherjee, Prasun K, Artemio Mendoza-Mendoza, Susanne Zeilinger, and Benjamin A Horwitz. 2022. "Mycoparasitism as a mechanism of *Trichoderma*-mediated suppression of plant diseases." *Fungal Biology Reviews* 39: 15-33.

Mukherjee, Prasun K, Aric Wiest, Nicolas Ruiz, Andrew Keightley, Maria E Moran-Diez, Kevin McCluskey, Yves François Pouchus, and Charles M Kenerley. 2011. "Two classes of new peptaibols are synthesized by a single non-ribosomal peptide synthetase of *Trichoderma* virens." *Journal of Biological Chemistry* 286 (6): 4544-4554.

Nadarajan, Stalin, and Surya Sukumaran. 2021. "Chemistry and toxicology behind chemical fertilizers." In *Controlled Release Fertilizers for Sustainable Agriculture*, 195-229. Elsevier.

Nagaraju, A, J Sudisha, S Mahadeva Murthy, and Shin-ichi Ito. 2012. "Seed priming with *Trichoderma harzianum* isolates enhances plant growth and induces resistance against *Plasmopara halstedii*, an incitant of sunflower downy mildew disease." *Australasian Plant Pathology* 41 (6): 609-620.

Nakkeeran, Sevugapperumal, Suppaiah Rajamanickam, Murugavel Vanthana, Perumal Renukadevi, and Malaiyandi Muthamilan. 2020. "Harnessing the Perception of *Trichoderma* Signal Molecules in Rhizosphere to Improve Soil Health and Plant Health." In *Trichoderma*, 61-79. Springer.

Nasution, Lita, Riahta Corah, Nuraida Nuraida, and Ameilia Zuliyanti Siregar. 2018. "Effectiveness *Trichoderma* and *Beauveria bassiana* on larvae of Oryctes rhinoceros on palm oil plant (*Elaeis guineensis* Jacq.) in vitro." *International Journal of Environment, Agriculture and Biotechnology* 3 (1): 239050.

Newman, SE, WM Brown, and N Ozbay. 2002. "The effect of the *Trichoderma harzianum* strains on the growth of tomato seedlings." XXVI *International Horticultural Congress: Managing Soil-Borne Pathogens: A Sound Rhizosphere to Improve Productivity in 635*.

Olsson, Lisbeth, Tove MIE Christensen, Kim P Hansen, and Eva A Palmqvist. 2003. "Influence of the carbon source on production of cellulases, hemicellulases and pectinases by *Trichoderma* reesei Rut C-30." *Enzyme and Microbial Technology* 33 (5): 612-619.

Omann, Markus, and Susanne Zeilinger. 2010. "How a mycoparasite employs G-protein signaling: using the example of *Trichoderma*." *Journal of Signal Transduction* 2010.

Ozbay, Nusret, and Steven E Newman. 2004. "Biological control with *Trichoderma* spp. with emphasis on *T. harzianum*." *Pakistan Journal of Biological Sciences* 7 (4): 478-484.

Pal, Kamal Krishna, and Brian McSpadden Gardener. 2006. "*Biological control of plant pathogens*."

Pascale, A, F Vinale, G Manganiello, M Nigro, S Lanzuise, M Ruocco, R Marra, N Lombardi, SL Woo, and M Lorito. 2017. "*Trichoderma* and its secondary metabolites improve yield and quality of grapes." *Crop protection* 92: 176-181.

Peterson, Robyn, and Helena Nevalainen. 2012. "*Trichoderma reesei* RUT-C30–thirty years of strain improvement." *Microbiology* 158 (1): 58-68.

Pieterse, Corné MJ, Christos Zamioudis, Roeland L Berendsen, David M Weller, Saskia CM Van Wees, and Peter AHM Bakker. 2014. "Induced systemic resistance by beneficial microbes." *Annual review of phytopathology* 52: 347-375.

Poveda, Jorge, Patricia Abril-Urias, and Carolina Escobar. 2020. "Biological control of plant-parasitic nematodes by filamentous fungi inducers of resistance: *Trichoderma*, mycorrhizal and endophytic fungi." *Frontiers in Microbiology* 11: 992.

Prajapati, Satyadev, Naresh Kumar, Sunil Kumar, and Shivam Maurya. 2020. "Biological control a sustainable approach for plant diseases management: A review." *Journal of Pharmacognosy Phytochemistry Reviews* 9 (2): 1514-1523.

Puyam, Anita. 2016. "Advent of *Trichoderma* as a bio-control agent-a review." *Journal of Applied and Natural Science* 8 (2): 1100-1109.

Rahimi Tamandegani, Parisa, Tamás Marik, Doustmorad Zafari, Dóra Balázs, Csaba Vágvölgyi, András Szekeres, and László Kredics. 2020. "Changes in peptaibol production of *Trichoderma* species during in vitro antagonistic interactions with fungal plant pathogens." *Biomolecules* 10 (5): 730.

Rani, Kavita, and Geeta Dhania. 2014. "Bioremediation and biodegradation of pesticide from contaminated soil and water—a noval approach." *Int J Curr Microbiol App Sci* 3 (10): 23-33.

Rao, Yuxin, Linzhou Zeng, Hong Jiang, Li Mei, and Yongjun Wang. 2022. "*Trichoderma atroviride* LZ42 releases volatile organic compounds promoting plant growth and suppressing *Fusarium wilt* disease in tomato seedlings." *BMC microbiology* 22 (1): 1-12.

Raupach, Georg S, and Joseph W Kloepper. 1998. "Mixtures of plant growth-promoting rhizobacteria enhance biological control of multiple cucumber pathogens." *Phytopathology* 88 (11): 1158-1164.

Reino, José Luis, Raul F Guerrero, Rosario Hernández-Galán, and Isidro G Collado. 2008. "Secondary metabolites from species of the biocontrol agent *Trichoderma*." *Phytochemistry Reviews* 7 (1): 89-123.

Rodríguez-González, Álvaro, Guzmán Carro-Huerga, Sara Mayo-Prieto, Alicia Lorenzana, Santiago Gutiérrez, Horacio J Peláez, and Pedro A Casquero. 2018. "Investigations of *Trichoderma* spp. and *Beauveria bassiana* as biological control agent for *Xylotrechus arvicola*, a major insect pest in Spanish vineyards." *Journal of economic entomology* 111 (6): 2585-2591.

Rodríguez-González, Álvaro, Pedro Antonio Casquero, Víctor Suárez-Villanueva, Guzmán Carro-Huerga, Samuel Álvarez-García, Sara Mayo-Prieto, Alicia Lorenzana, Rosa Elena Cardoza, and Santiago Gutiérrez. 2018. "Effect of trichodiene production by *Trichoderma harzianum* on *Acanthoscelides obtectus*." *Journal of Stored Products Research* 77: 231-239.

Romano, Stefano, Stephen A Jackson, Sloane Patry, and Alan DW Dobson. 2018. "Extending the "one strain many compounds"(OSMAC) principle to marine microorganisms." *Marine drugs* 16 (7): 244.

Rousseau, Annie, Nicole Benhamou, Ilan Chet, and Yves Piché. 1996. "Mycoparasitism of the extramatrical phase of Glomus intraradices by *Trichoderma harzianum*." *Phytopathology* 86 (5): 434-443.

Roy, Madhumita, Ashok K Giri, Sourav Dutta, and Pritam Mukherjee. 2015. "Integrated phytobial remediation for sustainable management of arsenic in soil and water." *Environment international* 75: 180-198.

Sachdev, Swati, and Rana Pratap Singh. 2020. "*Trichoderma*: a multifaceted fungus for sustainable agriculture." In *Ecological and practical applications for sustainable agriculture*, 261-304. Springer.

Sadoma, MT, ABB El-Sayed, and SM El-Moghazy. 2011. "Biological control of downy mildew disease of maize caused by *Peronosclerospora sorghi* using certain biocontrol agents alone or in combination." *J. Agric. Res* 37 (1): 1-11.

Sala, Arnau, Silvana Vittone, Raquel Barrena, Antoni Sanchez, and Adriana Artola. 2021. "Scanning agro-industrial wastes as substrates for fungal biopesticide production: Use of *Beauveria bassiana* and *Trichoderma harzianum* in solid-state fermentation." *Journal of Environmental Management* 295: 113113.

Salas-Marina, Miguel Angel, Miguel Angel Silva-Flores, Edith Elena Uresti-Rivera, Ernestina Castro-Longoria, Alfredo Herrera-Estrella, and Sergio Casas-Flores. 2011. "Colonization of Arabidopsis roots by *Trichoderma atroviride* promotes growth and enhances systemic disease resistance through jasmonic acid/ethylene and salicylic acid pathways." *European Journal of Plant Pathology* 131 (1): 15-26.

Sallam, Nashwa, Amal MI Eraky, and Ahmed Sallam. 2019. "Effect of *Trichoderma* spp. on *Fusarium wilt* disease of tomato." *Molecular biology reports* 46 (4): 4463-4470.

Sanjeev, K, and A Eswaran. 2008. "Efficacy of Micro Nutrients on Banana fusarium Wilt.(*Fusarium oxysporum* f. sp. cubense) and it's Synergistic Action with *Trichoderma viride*." *Notulae Botanicae Horti Agrobotanici Cluj-Napoca* 36 (1): 52-54.

Sanjeev, Kumar, Thakur Manibhushan, and RANI Archana. 2014. "*Trichoderma*: Mass production, formulation, quality control, delivery and its scope in commercialization in India for the management of plant diseases." *African journal of agricultural research* 9 (53): 3838-3852.

Saravanakumar, Kandasamy, Lili Fan, Kehe Fu, Chuanjin Yu, Meng Wang, Hai Xia, Jianan Sun, Yaqian Li, and Jie Chen. 2016. "Cellulase from *Trichoderma harzianum* interacts with roots and triggers induced systemic resistance to foliar disease in maize." *Scientific reports* 6 (1): 1-18.

Sarma, Birinchi K, Sudheer K Yadav, Jai S Patel, and Harikesh B Singh. 2014. "Molecular mechanisms of interactions of *Trichoderma* with other fungal species." *Open Mycol J* 8: 140-147.

Savary, Serge, Andrea Ficke, Jean-Noël Aubertot, and Clayton Hollier. 2012. "Crop losses due to diseases and their implications for global food production losses and food security." *Food security* 4 (4): 519-537.

Scarselletti, R, and JL Faull. 1994. "In vitro activity of 6-pentyl-α-pyrone, a metabolite of *Trichoderma harzianum*, in the inhibition of *Rhizoctonia solani* and *Fusarium oxysporum* f. sp. lycopersici." *Mycological Research* 98 (10): 1207-1209.

Schuster, André, and Monika Schmoll. 2010. "Biology and biotechnology of *Trichoderma*." *Applied microbiology and biotechnology* 87 (3): 787-799.

Seiboth, Bernhard, Christa Ivanova, Verena Seidl-Seiboth, and prospects. 2011. "*Trichoderma reesei*: a fungal enzyme producer for cellulosic biofuels." *Biofuel production-recent developments*: 309-340.

Sharfman, Miya, Maya Bar, Silvia Schuster, Meirav Leibman, and Adi Avni. 2014. "Sterol-dependent induction of plant defense responses by a microbe-associated molecular pattern from *Trichoderma viride*." *Plant Physiology* 164 (2): 819-827.

Sharma, Harsh P, Hiral Patel, and Sugandha. 2017. "Enzymatic added extraction and clarification of fruit juices–A review." *Critical reviews in food science and nutrition* 57 (6): 1215-1227.

Sharma, Prashant Kumar, and R Gothalwal. 2017. "*Trichoderma*: a potent fungus as biological control agent." In *Agro-environmental sustainability*, 113-125. Springer.

Sharma, Sushma, Divjot Kour, Kusam Lata Rana, Anu Dhiman, Shiwani Thakur, Priyanka Thakur, Sapna Thakur, Neelam Thakur, Surya Sudheer, and Neelam Yadav. 2019. "*Trichoderma*: biodiversity, ecological significances, and industrial applications." In *Recent advancement in white biotechnology through fungi*, 85-120. Springer.

Sharma, Vivek, Richa Salwan, PN Sharma, and Arvind Gulati. 2017. "Integrated translatome and proteome: approach for accurate portraying of widespread multifunctional aspects of *Trichoderma*." *Frontiers in Microbiology* 8: 1602.

Sharon, Edna, Ilan Chet, and Yitzhak Spiegel. 2011. "*Trichoderma* as a biological control agent." In *Biological Control of Plant-Parasitic Nematodes:*, 183-201. Springer.

Shoresh, Michal, Gary E Harman, and Fatemeh Mastouri. 2010. "Induced systemic resistance and plant responses to fungal biocontrol agents." *Annual review of phytopathology* 48 (1): 21-43.

Shukla, Nandani, RP Awasthi, Laxmi Rawat, and J Kumar. 2012. "Biochemical and physiological responses of rice (*Oryza sativa* L.) as influenced by *Trichoderma* harzianum under drought stress." *Plant Physiology Biochemistry* 54: 78-88.

Silva, Roberto N, Valdirene Neves Monteiro, Andrei Stecca Steindorff, Eriston Vieira Gomes, Eliane Ferreira Noronha, and Cirano J Ulhoa. 2019. "*Trichoderma*/pathogen/ plant interaction in pre-harvest food security." *Fungal biology* 123 (8): 565-583.

Singh, Akansha, Nandini Shukla, BC Kabadwal, AK Tewari, and J Kumar. 2018a. "Review on plant-*Trichoderma*-pathogen interaction." *International Journal of Current Microbiology and Applied Sciences* 7 (2): 2382-2397.

Singh, Akansha, Nandini Shukla, BC Kabadwal, AK Tewari, and J Kumar. 2018b. "Review on plant-*Trichoderma*-pathogen interaction." *International Journal of Current Microbiology Applied Sciences* 7 (2): 2382-2397.

Sivasithamparamb, Krishnapillai, Emilio L Ghisalbertic, Roberta Marraa, Sheridan L Wooa, and Matteo Loritoa. 2008. "*Trichoderma*–plant–pathogen interactions." *Soil Biology Biochemistry* 40: 1-10.

Soccol, Carlos Ricardo, Vincenza Faraco, Susan G Karp, Luciana PS Vandenberghe, Vanete Thomaz-Soccol, Adenise L Woiciechowski, and Ashok Pandey. 2019. "Lignocellulosic bioethanol: current status and future perspectives." *Biofuels: alternative feedstocks and conversion processes for the production of liquid and gaseous biofuels*: 331-354.

Sood, Monika, Dhriti Kapoor, Vipul Kumar, Mohamed S Sheteiwy, Muthusamy Ramakrishnan, Marco Landi, Fabrizio Araniti, and Anket Sharma. 2020. "*Trichoderma*: The "secrets" of a multitalented biocontrol agent." *Plants* 9 (6): 762.

Spaepen, Stijn. 2015. "Plant hormones produced by microbes." In *Principles of plant-microbe interactions*, 247-256. Springer.

Srinon, W, K Chuncheen, K Jirattiwarutkul, K Soytong, and S Kanokmedhakul. 2006. "Efficacies of antagonistic fungi against *Fusarium wilt* disease of cucumber and tomato and the assay of its enzyme activity." *Journal of Agricultural Technology* 2 (2): 191-201.

Steyaert, JM, HJ Ridgway, Y Elad, and A Stewart. 2003. "Genetic basis of mycoparasitism: a mechanism of biological control by species of *Trichoderma*." *New Zealand Journal of Crop and Horticultural Science* 31 (4): 281-291.

Steyaert, Johanna M, Richard J Weld, Artemio Mendoza-Mendoza, S Kryštofová, M Šimkovič, L Varečka, and Alison Stewart. 2013. "Asexual development in *Trichoderma*: from conidia to chlamydospores." *J Trichoderma: Biology Applications*: 87-109.

Suzuki, Nobuhiro, Rosa M Rivero, Vladimir Shulaev, Eduardo Blumwald, and Ron Mittler. 2014. "Abiotic and biotic stress combinations." *New Phytologist* 203 (1): 32-43.

Swain, Harekrushna, and Arup K Mukherjee. 2020. "Host–pathogen–*Trichoderma* interaction." In *Trichoderma*, 149-165. Springer.

Syamala D., Nabanita Kumar S. and Lalitha P. . "Mitigation of Aflatoxin Contamination in Groundnuts using *Trichoderma viride*." *Res. J. Chem. Environ* 25 (12): 32-43. https://doi.org/https://doi.org/10.25303/2512rjce3243;.

Tchameni, Severin Nguemezi, Mihaela Cotârleț, Ioana Otilia Ghinea, Marie Ampere Boat Bedine, Modeste Lambert Sameza, Daniela Borda, Gabriela Bahrim, and Rodica Mihaela Dinică. 2020. "Involvement of lytic enzymes and secondary metabolites produced by *Trichoderma* spp. in the biological control of *Pythium myriotylum*." *International Microbiology* 23 (2): 179-188.

Tondje, PR, DP Roberts, MARIE-CLAUDE Bon, TOMOTHY Widmer, GJ Samuels, A Ismaiel, AD Begoude, T Tchana, E Nyemb-Tshomb, and M Ndoumbe-Nkeng. 2007. "Isolation and identification of mycoparasitic isolates of *Trichoderma asperellum* with potential for suppression of black pod disease of cacao in Cameroon." *Biological control* 43 (2): 202-212.

Tortella, Gonzalo R, Maria Cristina Diez, and Nelson Durán. 2005. "Fungal diversity and use in decomposition of environmental pollutants." *Critical reviews in microbiology* 31 (4): 197-212.

Trillas, M Isabel, Eva Casanova, Lurdes Cotxarrera, José Ordovás, Celia Borrero, and Manuel Avilés. 2006. "Composts from agricultural waste and the *Trichoderma asperellum* strain T-34 suppress *Rhizoctonia solani* in cucumber seedlings." *Biological control* 39 (1): 32-38.

Tripathi, Pratibha, Poonam C Singh, Aradhana Mishra, Puneet S Chauhan, Sanjay Dwivedi, Ritu Thakur Bais, and Rudra Deo Tripathi. 2013a. "*Trichoderma*: a potential bioremediator for environmental clean up." *Clean Technologies and Environmental Policy* 15 (4): 541-550.

Tripathi, Pratibha, Poonam C Singh, Aradhana Mishra, Puneet S Chauhan, Sanjay Dwivedi, Ritu Thakur Bais, and Rudra Deo Tripathi. 2013b. "*Trichoderma*: a potential bioremediator for environmental clean up." *Clean Technologies Environmental Policy* 15 (4): 541-550.

Tucci, Marina, Michelina Ruocco, Luigi De Masi, Monica De Palma, and Matteo Lorito. 2011. "The beneficial effect of *Trichoderma* spp. on tomato is modulated by the plant genotype." *Molecular plant pathology* 12 (4): 341-354.

Tudi, Muyesaier, Huada Daniel Ruan, Li Wang, Jia Lyu, Ross Sadler, Des Connell, Cordia Chu, and Dung Tri Phung. 2021. "*Agriculture Development, Pesticide Application and Its Impact on the Environment*." 18 (3): 1112. https://www.mdpi. com/1660-4601/18/3/1112.

Tyśkiewicz, Renata, Artur Nowak, Ewa Ozimek, and Jolanta Jaroszuk-Ściseł. 2022. "*Trichoderma*: The current status of its application in agriculture for the biocontrol of fungal phytopathogens and stimulation of plant growth." *International Journal of Molecular Sciences* 23 (4): 2329.

Varsha, M, P Senthil Kumar, and B Senthil Rathi. 2022. "A review on recent trends in the removal of emerging contaminants from aquatic environment using low-cost adsorbents." *Chemosphere* 287: 132270.

Verma, Mausam, Satinder K Brar, RD Tyagi, R nY Surampalli, and JR Valero. 2007. "Antagonistic fungi, *Trichoderma* spp.: panoply of biological control." *Biochemical Engineering Journal* 37 (1): 1-20.

Viera, William, Michelle Noboa, Aníbal Martínez, Francisco Báez, Rosendo Jácome, Lorena Medina, and Trevor Jackson. 2019. "*Trichoderma asperellum* increases crop yield and fruit weight of blackberry (*Rubus glaucus*) under subtropical Andean conditions." *Vegetos* 32 (2): 209-215.

Villalobos-Escobedo, José M, Saraí Esparza-Reynoso, Ramón Pelagio-Flores, Fabiola López-Ramírez, León F Ruiz-Herrera, José López-Bucio, and Alfredo Herrera-Estrella. 2020. "The fungal NADPH oxidase is an essential element for the molecular dialog between *Trichoderma* and Arabidopsis." *The Plant Journal* 103 (6): 2178-2192.

Vinale, Francesco, Krishnapillai Sivasithamparam, Emilio L Ghisalberti, Roberta Marra, Sheridan L Woo, and Matteo Lorito. 2008. "*Trichoderma*–plant–pathogen interactions." *Soil Biology Biochemistry* 40 (1): 1-10.

Vinale, Francesco, Krishnapillai Sivasithamparam, Emilio L Ghisalberti, Michelina Ruocco, Sheridan Woo, and Matteo Lorito. 2012. "*Trichoderma* secondary metabolites that affect plant metabolism." *Natural product communications* 7 (11): 1934578X1200701133.

Wallenstein, Matthew D, and Richard G Burns. 2011. "Ecology of extracellular enzyme activities and organic matter degradation in soil: A complex community-driven process." *Methods of soil enzymology* 9: 35-55.

Wong, Ken KY, and John N Saddler. 1992. "*Trichoderma* xylanases, their properties and application." *Critical Reviews in Biotechnology* 12 (5-6): 413-435.

Woo, Sheridan L, Michelina Ruocco, Francesco Vinale, Marco Nigro, Roberta Marra, Nadia Lombardi, Alberto Pascale, Stefania Lanzuise, Gelsomina Manganiello, and Matteo Lorito. 2014. "*Trichoderma*-based products and their widespread use in agriculture." *The Open Mycology Journal* 8 (1).

Yaashikaa, PR, P Senthil Kumar, A Saravanan, and Dai-Viet N Vo. 2021. "Advances in biosorbents for removal of environmental pollutants: A review on pretreatment, removal mechanism and future outlook." *Journal of Hazardous Materials* 420: 126596.

Yacob, Norhidayah. 2019. "Antagonistic Activity of *Trichoderma parareesei* and *Trichoderma harzianum* against *Colletotrichum* sp., a Causal Pathogen for Anthracnose Disease in Chilli." University Malaysia Kelantan.

Zafra, German, and Diana V Cortés-Espinosa. 2015. "Biodegradation of polycyclic aromatic hydrocarbons by *Trichoderma* species: a mini review." *Environmental Science and Pollution Research* 22 (24): 19426-19433.

Zahran, Zulaikha, Nik Mohd Izham Mohamed Nor, Hamady Dieng, Tomomitsu Satho, and Abdul Hafiz Ab Majid. 2017. "Laboratory efficacy of mycoparasitic fungi (*Aspergillus tubingensis* and *Trichoderma harzianum*) against tropical bed bugs (*Cimex hemipterus*) (Hemiptera: Cimicidae)." *Asian Pacific Journal of Tropical Biomedicine* 7 (4): 288-293.

Zeilinger, Susanne, and Markus Omann. 2007. "*Trichoderma* biocontrol: signal transduction pathways involved in host sensing and mycoparasitism." *Gene regulation and systems biology* 1: GRSB. S397.

Zhang, Yi, and Wen-Ying Zhuang. 2022. "MAPK Cascades Mediating Biocontrol Activity of *Trichoderma* brevicrassum Strain TC967." *Journal of Agricultural and Food Chemistry* 70 (8): 2762-2775.

Chapter 2

Trichoderma Sp: Bioproducts and Their Main Uses in Agriculture

Francisco Wilson Reichert Júnior[1,*]
Jéssica Mulinari[2]
Aline Frumi Camargo[3,4]
Thamarys Scapini[5]
José Luís Trevizan Chiomento[1]
Eduardo José Pedroso Pritsch[3]
Caroline Berto[3]
Laura Helena dos Santos[3]
Gislaine Fongaro[4]
Altemir José Mossi[3]
and Helen Treichel[3,4,†]

[1]University of Passo Fundo, School of Agricultural Sciences, Innovation and Business, Passo Fundo, Brazil
[2]Federal University of Santa Catarina, Department of Chemical Engineering and Food Engineering, Florianópolis, Brazil
[3]Federal University of Fronteira Sul, Laboratory of Microbiology and Bioprocesses, Erechim, Brazil
[4]Federal University of Santa Catarina, Graduate Program in Biotechnology and Biosciences, Florianópolis, Brazil
[5]Federal University of Paraná, Department of Bioprocess Engineering and Biotechnology, Curitiba, Brazil

† Corresponding Author's Email: helentreichel@gmail.com.

In: Trichoderma: Taxonomy, Biodiversity and Applications
Editor: Michael S. Mouton
ISBN: 979-8-88697-946-6

Abstract

The current agricultural model, based heavily on the chemical management of pests and diseases, generates several environmental and public health problems. In this sense, searching for alternatives to chemical control is essential to make agriculture more ecologically sustainable. Fungi of the genus *Trichoderma* have been widely studied as a biological control of pests and diseases that affect crops. Bioproducts based on *Trichoderma* spp. have been used in about 6 million hectares in Brazil, mainly to control diseases caused by soil fungi (*Fusarium, Rhizoctonia, Sclerotinia, Verticillium, Phytophthora, Pythium, Armillaria,* and *Roselinia*) in beans, soybeans, corn, strawberries, among others. Some products are also recommended for controlling pathogens that cause lesions and rot in leaves, branches, stems, fruits, and flowers, such as Botrytis cinerea, which causes gray mold on strawberry and ornamental plants. In addition to biological control, although little has been studied in Brazil so far, these fungi can increase the growth and productivity of several plants. This effect responds to the modifications they cause in root architecture, leading to more efficient water use and more significant acquisition of mineral nutrients. Therefore, this chapter aims to present the main bioproducts derived from *Trichoderma* and the primary benefits of these microorganisms in agriculture. For this, the chapter addresses the following topics: (i) Where to find *Trichoderma* species; (ii) *Trichoderma*-based products; (iii) *Trichoderma* for phytopathogenic fungi control; (iv) *Trichoderma* for weed control and (v) *Trichoderma* as a plant growth promoter.

Keywords: biological control, biopesticide, sustainable agriculture, biofertilizer

Where to Find *Trichoderma* Species

Trichoderma is a fungus of the Hypocreaceae family that has attracted considerable attention since the 1970s. There are more than 10,000 species found in nature, divided into five different taxonomic sections: *Saturnisporum*, *Pachybasium*, *Longibrahiatum*, *Trichoderma*, and *Hypocreanum* (Zin and Badaluddin, 2020). With an optimum growth temperature of 25-30°C (Sandle, 2014), these species exist in diverse habitats, such as natural soils, rotten wood, living plants, mushroom substrates, the human body, and agricultural environments (Bai et al., 2022). They are among the most disseminated fungi species globally, found in ecosystems ranging from tundra to the tropics (Samuels, 1996; Silva et al., 2014). They are essential to the ecosystem's balance, are considered the principal decomposers in the ecosphere, and are often the main components of the mycoflora in various soils (Felix et al., 2014).

To evolve efficiently, *Trichoderma* fungi developed a fine metabolic regulation capable of adjusting to different environmental conditions (Silva et al., 2014). Some species, for example, can change primary and secondary metabolic pathways according to oxygen availability in the environment, surviving oxygen scarcity by using fermentation metabolism (Chovanec et al., 2005). They can also produce antibiotics or toxins and establish ways to compete for nutrients with other microorganisms (Benítez et al., 2004). Moreover, some *Trichoderma* can degrade toxic compounds, such as pesticides, and survive in environments with high concentrations of metals (Errasquín and Vázquez, 2003; Katayama et al., 2001).

Jiang et al. (2016) analyzed the consequence of seasonal, regional, and crop variability for haplotype biodiversity in the main agricultural areas of China. They also assembled a potential distribution pattern of *Trichoderma* species in the country. Two thousand seventy-eight samples were collected during three seasons of the year; the amounts of *Trichoderma* isolated in spring were sharply lower than in summer and autumn. Throughout the study, it was observed that the composition and occurrence of species depended on the region and the types of cultivation applied. Furthermore, significant *Trichoderma* communities were found in different cropping soils. The design and proportion of *Trichoderma* species varied according to the four other provinces studied (Zhejiang, Jiangsu, Anhui, and Shandong). However, the species *T. harzianum*, *T. asperellum*, *T. hamatum*, and *T. virens* were identified as dominant.

The occurrence of Trichoderma in different ecological niches (see Table 1) reflects a significant genetic and metabolic variability among species, enabling diverse applications and technological innovations from them (Jiang et al., 2016). The ability to survive different environmental conditions in nature and attack other fungal species has inspired several research over the years, stimulating applications for Trichoderma, mainly in biological control in agriculture (Bai et al., 2022; Olowe et al., 2022). The following section addresses the use of *Trichoderma* and its byproducts (secondary metabolites and enzymes) in biocontrol, biofertilization, and soil remediation, among others.

Trichoderma-Based Products

Since the pioneering research on *Trichoderma* species more than 50 years ago, the expanding applications of these microorganisms have gained more attention each year (Dutta et al., 2022; Paymaneh et al., 2023). The thematic map (Figure 1) was generated from the compilation of recent research papers (2020-2021) aiming to highlight the main applications currently given to *Trichoderma* species. Applications in agriculture (e.g., biological control, biocontrol, biofertilizer) are the primary uses of this species, which can be noted by the presence of different clusters on the thematic map that present incidences of terms, such as antifungal activity, plant growth promotion, mycoparasitism, and others.

The applications of *Trichoderma* species in agriculture are mainly linked to the metabolism of these microorganisms, which present high resistance to several pesticides (Camargo et al., 2019; Mostafa et al., 2022), the potential for mycoparasitism and antibiosis (Dutta et al., 2022), high production of extracellular enzymes and secondary metabolites (Bordin et al., 2021; Cavalcante et al., 2021; Taylor et al., 2020; Ulrich et al., 2021), besides presenting an improved capacity to provide nutrients in the soil (Bellini et al., 2023; Gilardi et al., 2019; Hang et al., 2022; Tančić-Živanov et al., 2020). Because of these characteristics, several species of the fungus *Trichoderma* are evaluated and used as agents of biological and integrated control, as well as in the growth promotion and defense of target plants for agriculture.

Besides applications in agriculture, some *Trichoderma* species were tested as entomopathogenic fungi against human disease vectors. Extracellular secondary metabolites of many fungi, including *Trichoderma*, have been studied as larvicides for many mosquito species as an alternative to chemical pesticides. Perera et al. (2023) used extracellular extracts of *T. longibrachiatum* and *T. viride* as larvicides against dengue vectors (*Aedes aegypti* and *Aedes albopictus*) in Sri Lanka. The biological control ability of *Trichoderma* spp. is mainly due to the production of lytic enzymes, such as proteases, N-acetylglucosaminidase, β-1,3-glucanase, and chitinases, which can destroy cell components (da Silveira et al., 2021; Kullnig et al., 2001; Shakeri and Foster, 2007). In *Aedes aegypti* larvae, an enzyme cocktail (N-acetylglucosaminidase, β-1,3-glucanase, and chitinases) produced by the fungus *T. asperellum* caused morphological changes in the larval body and rupture of the head and thorax, with marked degradation of the cuticle

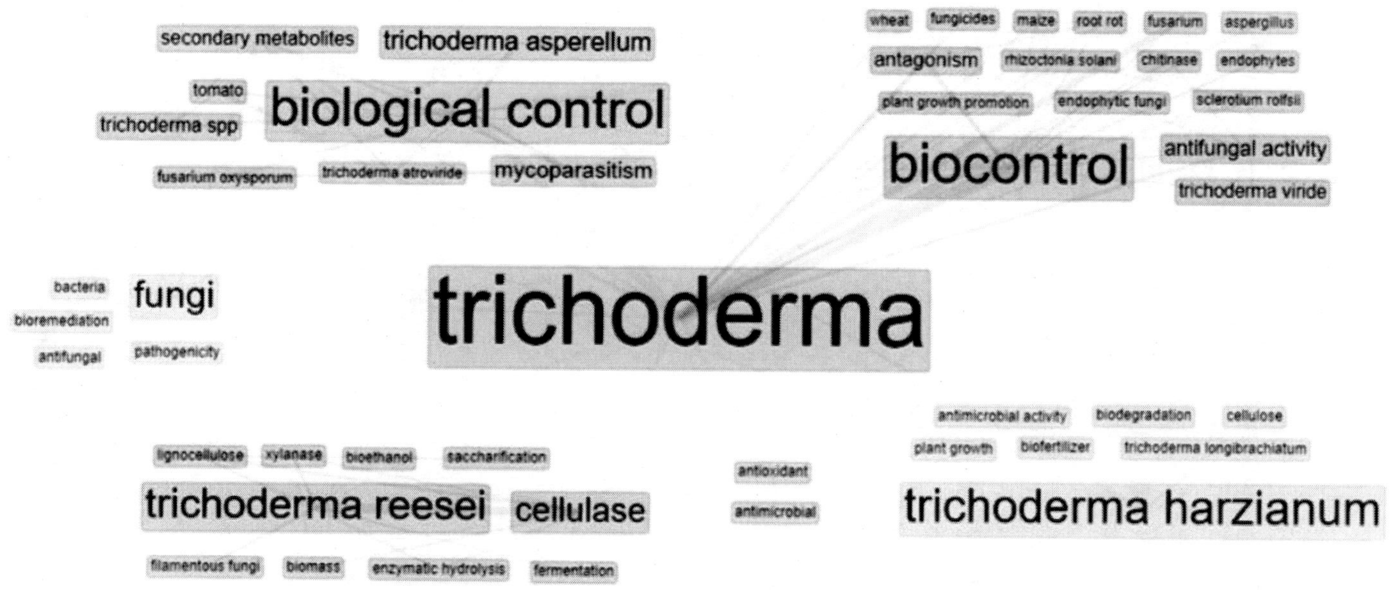

Figure 1. Thematic map made from 2,000 research papers published between 2020 and 2021, using the keywords "*Trichoderma*" limited to "English" and "Journal" from data on the Scopus platform using Bibliometrix (Aria and Cuccurullo, 2017).

surface (da Silveira et al., 2021). The enzymatic process facilitates the penetration of fungal structures into the larval cuticle, with subsequent vegetative fungal growth (Podder and Ghosh, 2019). In addition, it has been reported that mechanical processes can act in synergism with enzyme production, as demonstrated by Podder and Ghose (2019), who isolated the fungus *T. asperellum* from the soil and tested it as a larvicide against the Anopheles mosquito (malaria vector); they observed a conidial blockage of the larval spiracles and mouth, suggesting that this blockage may be the probable cause of death of the larvae. *T. asperellum* is well-known for its ability to produce enzyme complexes capable of interrupting cuticle formation in the larval stage (da Silveira et al., 2021; Podder and Ghosh, 2019; Sebumpan et al., 2022).

Soil bioremediation is also a field where *Trichoderma* can be applied, mainly to metabolize residual compounds associated with conventional agriculture (e.g., pesticides) and heavy metals. In a recent study, *Trichoderma* isolated from Amazonian soil demonstrated the ability to grow in the presence of glyphosate as a source of phosphorus (Correa et al., 2021). *T. viride* isolated from contaminated soil demonstrated tolerance to 400 ppm of diazinon pesticide and a degradation capacity of 80.26% (Mostafa et al., 2022). Several mechanisms may be related to the bioremediation capacity of these compounds, suggesting the involvement of processes of alkylation, leaching, methylation, and redox transformations, in addition to the secretion of several extracellular enzymes that can degrade hazardous pollutants and transform them into non-toxic forms (Bilal and Iqbal, 2020; Mostafa et al., 2022; Saravanan et al., 2021). Furthermore, specific genes can be critical for degradation in bioremediation processes, as discussed by Malmir et al. (2022), which identified the presence of the cyanide hydratase enzyme gene in *T. harzianum*, a key enzyme for cyanide degradation. The microorganism degraded 75% of the cyanide compound at a concentration of 15 mM in 7 days (Malmir et al., 2022). Regarding the bioremediation of heavy metals, *Trichoderma* fungi present promising results compared with already consolidated processes (e.g., bacteria and chemical processes) (Dell'Anno et al., 2022). The ability to bioremediate chemical compounds utilized in conventional agriculture has been associated with the high capacity of *Trichoderma* to tolerate these compounds and other bioavailable nutrients in the soil (Escudero-Leyva et al., 2022). These strategies can enable the development of biobased bioproducts with the *Trichoderma* fungus in the integrated management of pests and diseases.

Not unexpectedly, one of the Clusters (purple) in Figure 1 highlights terms such as *Trichoderma reesei* and cellulase. Since the isolation of the *T. reesei* fungus, its characteristic to degrade and grow in cellulose-rich environments has been explored, pioneering the concept of enzymatic saccharification of cellulose by a synergistic combination of cellulases (Bischof et al., 2016). The application of these enzymes is strongly dependent on the development of biorefineries based on lignocellulosic biomass (e.g., sugarcane bagasse, agricultural residues) for conversion into biofuel - as can be seen in the purple cluster of Figure 1. These enzymes are mainly associated with biomass saccharification for ethanol production and, more recently, in developing chemical platform products (Fang et al., 2023; Shankar et al., 2022). Currently, the ability related to the secretion of enzymatic cocktails of cellulases is widely exploited by industries, such as Novozymes Inc., that uses *T. reesei* as a base strain, isolating cellulolytic and non-cellulolytic genes to transform the microorganism for the preparation of an efficient enzymatic cocktail for saccharification of lignocellulosic biomass (Adsul et al., 2020; Bischof et al., 2016). The fungus *T. reesei* is not the only one of the species with the capacity to produce cellulolytic enzymes; *Trichoderma* is recognized as one of the main filamentous fungi species for the production of enzymatic cocktails, such as cellulases, peroxidases, xylanases, chitinases, and proteases (Adsul et al., 2020; Bonatto et al., 2023). For example, *T. citrinoviride* MDU-1 and *T. longibrachiatum* MDU-6 can release sugar in pretreated switchgrass biomass aiming at ethanol production (Saini et al., 2022).

The versatility of the enzymatic cocktails produced by the *Trichoderma* species has attracted the attention of several researchers. The adaptation to different substrates, the resistance to various compounds, and the metabolic versatility of these microorganisms make it possible to identify genes of interest or bioactive activity of these organisms in different fields beyond agriculture or biofuels. A recent strategy is the application of enzyme cocktails from *T. koningiopsis* produced from microalgae biomass. The extracted enzymes (mainly peroxidase) could decolorize dyes (Klanovicz et al., 2022). Similar results were discussed for *T. asperellum*, a producer of laccases capable of decolorizing synthetic dyes in a saline environment (Ali et al., 2020). In co-culture with other filamentous fungi, *T. longibrachiatum* was used for the removal of pharmaceutical drugs (carbamazepine, diclofenac sodium, ibuprofen), and the tolerance of the fungi to the presence of contaminants was positively correlated with enzymatic activity (mainly manganese peroxidase) (Kasonga et al., 2022).

Table 1. Overview of recent studies on the isolation of *Trichoderma* fungi from different environments

Trichoderma species	Isolation site	Application	Reference
T. asperellum	Agricultural soil	Plant growth promotion	(Singh et al., 2016)
T. asperellum	Soil samples	Control of plant fungal pathogens and flower-inducing ability	(Sharma et al., 2022)
T. asperellum	Semi-arid soil (Saudi Arabia)	Integrated diseases management and growth promotion	(El-Komy et al., 2022)
T. asperellum	Seawater environments (harbors)	Production of oxidative enzymes and dyes decolorization	(Ali et al., 2020)
T. asperellum *T. hamatum*	Marine sponges and corals	Fungicidal activity and rice yield improvement	(Klaram et al., 2022)
T. asperellum	Soil samples	Control of *Anopheles* spp. (malaria vectors)	(Podder and Ghosh, 2019)
T. atroviride 1133	Amazonian aquatic environments	Larvicidal activities against *Aedes aegypti*	(de Oliveira et al., 2021)
T. atroveride	Mangrove ecosystem	Dye decolorization and salt tolerance	(De Paula et al., 2022)
T. atroviride H548	Mangrove sediment	Antifungal activity	(Tang et al., 2021)
T. dorothopsis	Soil samples (rhizosphere of a bell or chili pepper plant)	Biocontrol (against *Phytophthora capsici*)	(Tomah et al., 2020)
T. gamsii M501	Forest soil	Enzymatic cocktail of xylanases and cellulases	(Baskaran and Krishnan, 2020)
T. gamsii	Common bean rhizosphere	Phosphate solubilization	(Bedine et al., 2022)
T. koningiopsis CQSQ1002	Fresh branches of the medicinal plant *Morinda officinalis*	Antibacterial and antitumor activities	(Chen et al., 2019)
T. koningiopsis MK860714	Infected plants (weed)	Bioherbicide (weed biocontrol)	(Reichert Júnior et al., 2019)
T. koningii *T. pseudokoningii*	Soil from Cerrado and Pantanal biomes (Brazil)	Xylanolytic enzymes in low-cost media (agro-industrial residues)	(Sanguine et al., 2022)
T. longibrachiatum	Areas surrounding Pretoria North, South Africa	Treatment of wastewater with pharmaceuticals drugs	(Kasonga et al., 2022)

The relevance of *Trichoderma* species for biotechnological development is unquestionable. Although the research is concentrated on the advances in agriculture with less dependence on agrochemicals, through the consolidated development of biocontrol, it is observed that the applications of the fungus *Trichoderma* have expanded more in other fields, as in the treatment of effluents and the development of enzymatic cocktails for saccharification of lignocellulosic biomass.

Phytopathogenic Fungi Control

Plant diseases are considered one of the leading causes of the reduction in the global-level production of food (Ghorbanpour et al., 2018). The most significant losses are attributed to soil-dwelling fungal phytopathogens (Almeida et al., 2019). Worldwide, more than 19,000 species of fungi are responsible for many crop diseases (Jain et al., 2019), most of which belong to the Ascomycota and Basidiomycota phyla (Doehlemann et al., 2017). Consequently, chemical fungicides are inappropriately used as the primary way of disease control, which forces phytopathogens to undergo genetic mutations attributed to the selection of fungicide-resistant biotypes (Sood et al., 2020). This fact, linked to the accumulation of harmful toxins in humans and ecosystems, encourages scientists, industry, and producers to invest in developing tools for alternative management of plant diseases, such as applying biological control agents.

About 90% of fungal biocontrol agents belong to different strains of *Trichoderma* (Hermosa et al., 2012); a microorganism considered an effective antagonist agent in the fight against disease-causing fungi for decades (Asad, 2022). *Trichoderma* feeds on nutrients from plant roots in the rhizosphere, establishes interactions, and targets the phytopathogens Rhizoctonia, Rhizopus, Endothia, Helminthosporium, Armillaria, Botrytis, Fusarium, and Pythium (Druzhinina et al., 2011). This fungus can be found in almost all types of soils under tropical and temperate environments (Adnan et al., 2019).

The literature reports that the beneficial properties of avirulent *Trichoderma* strains allow their use in plant protection. *Trichoderma* spp. uses multiple complex mechanisms against fungal phytopathogens, which interact together in the biocontrol phenomenon, depending on the antagonistic microorganism–phytopathogen–host plant interface (Ghorbanpour et al., 2018). The direct effects on plant pathogens include the synthesis of antibiotics, the production of cell wall degrading enzymes, the establishment of mycoparasitism, and competition for space and nutrients, mainly carbon, nitrogen, and iron (Druzhinina et al., 2011; Jaroszuk-Ściseł et al., 2019). Indirectly, *Trichoderma* induces local or systemic plant resistance through endoelictors (from cell walls of the host plant) and exoeliciters (from the infecting microorganism) (Saravanakumar et al., 2016). Table 2 shows several studies that reported that the performance of cultures against phytopathogens is improved when cultivated with *Trichoderma* spp.

Table 2. Use of *Trichoderma* spp. in the control of phytopathogens in crops

Crop	*Trichoderma* species	Phytopathogen	Reference
Apple (*Malus domestica* Borkh.)	*T. harzianum* T88 *T. atroviride* T95	*Botryosphaeria berengeriana* f. sp. *Piricola*	(Kexiang et al., 2002)
Banana (*Musa* spp.)	*T. piluliferum*	*Colletotrichum musae*	(da Costa et al., 2021)
Beans (*Phaseolus vulgaris* L.)	*T. viride* Ts-1	*Sclerotinia sclerotiorum*	(Amin et al., 2010)
Blueberry (*Vaccinium corymbosum* L.)	*T. atroviride* IC-11	*Botrytis cinerea*	(Bello et al., 2022)
Cacao (*Theobroma cacao* L.)	*T. martiale* ALF 247	*Phytophthora palmivora*	(Hanada et al., 2009)
Chickpea (*Cicer arietinum* L.)	*T. hamatum* CECT 20103 (Thm) *T. koningii* CECT 2936 (Tk)	*Ascochyta rabiei*	(Poveda, 2021)
Cucumber (*Cucumis sativus* L.)	*T. asperellum* 525 *T. harzianum* 610 *T. pseudokoningii* 886	*Fusarium oxysporum* f. sp. *cucumerinum*	(Li et al., 2019)
	T. harzianum *T. viride*	*Podosphaera xanthii*	(Sarhan et al., 2020)
Eggplant (*Solanum melongena* L.)	*T. harzianum*	*Macrophomina phaseolina*	(Ramezani, 2008)
Etlingera linguiformis (Roxb.) R.M.Sm	*T. harzianum*	*Curvularia lunata* var. *aeriai*	(Kithan and L, 2014)
Ginger (*Zingiber officinale* Rosc.)	*T. viride* *T. harzianum*	*F. oxysporum* f.sp. *zingiberi*	(Khatso and Ao, 2013)
Lettuce (*Lactuca sativa* L.)	*T. spirale* T76-1	*Corynespora cassiicola* *Curvularia aeria*	(Baiyee et al., 2019)
Olive (*Olea europaea* L.)	*T. asperellum* Bt3 and T25	*Verticillium dahliae*	(Carrero-Carrón et al., 2016)
Pepper (*Capsicum annuum* L.)	*T. hamatum* MHT1134	*F. oxysporum*	(Mao et al., 2020)
	T. koningii 7a and 7c	*Pythium ultimum*	(Harris, 1999)
Potato (*Solanum Tuberosum* L.)	*T. harzianum*	*Rhizoctonia solani*	(Pandey and Pundhir, 2013)
Soybean [*Glycine max* (L.) Merr.]	*T. koningiopsis*	*C. truncatum*	(Silva et al., 2020)

Crop	*Trichoderma* species	Phytopathogen	Reference
Strawberry (*Fragaria* X *ananassa* Duch.)	*T. atrobrunneum* T17	*Armillaria mellea*	(Rees et al., 2022)
	T. hamatum T-105	*C. acutatum*	(Freeman et al., 2004)
Tea [*Camellia sinensis* (L.) O. Kuntze]	*T. viride* SDRLIN1	*Pestalotia theae*	(Naglot et al., 2015)
Tobacco (*Nicotiana tabacum* L.)	*T. harzianum*	*Alternaria alternata*	(Gveroska and Ziberoski, 2012)
Tomato (*Solanum lycopersicum* L.)	*T. harzianum* TM	*B. cinerea*	(Geng et al., 2022)
Wheat (*Triticum aestivum* L.)	*T. viride*	*Ustilago segetum*	(Singh, 2004)

Table 3. Interactions between *Trichoderma* spp. and other beneficial microbes in controlling phytopathogens in crops

Crop	Interactions	Phytopathogen	Reference
Cacao (*T. cacao* L.)	*T. asperellum*, *Gigaspora margarita* and *Acaulospora tuberculata*	*Phytophthora megakarya*	(Tchameni et al., 2011)
Jerusalem artichoke (*Helianthus tuberosus* L.)	*T. harzianum* and *Glomus clarum*	*Sclerotium rolfsii*	(Sennoi et al., 2013)
Melon (*Cucumis melo* L.)	*T. harzianum*, *Glomus intraradices* and *Glomus mosseae*	*F. oxysporum*	(Martínez-Medina et al., 2011)
Physic nut (*Jatropha curcas* L.)	*T. viride*, *Pseudomonas fluorescens* and *Bacillus subtilis*	*Lasiodiplodia theobromae*	(Latha et al., 2011)
Potato (*Solanum tuberosum* L.)	*T. harzianum* and *B. subtilis*	*Streptomyces* spp.	(Wang et al., 2019)
Soybean [*G. max* (L.) Merr.]	*T. harzianum* and *Pseudomonas fluorescens*	*Rhizoctonia solani*, *Sclerotium rolfsii*, and *Macrophomina phaseolina*	(Mishra et al., 2011)
Tobacco (*N. tabacum* L.)	*T. harzianum* and *G. mosseae*	*Ralstonia solanacearum*	(Yuan et al., 2016)
Tomato (*S. lycopersicum* L.)	*T. harzianum*, *Acaulospora*, *Gigaspora* and *Glomus*	*F. oxysporum* f. sp. *Lycopersici*	(Mwangi et al., 2011)
Wheat (*T. aestivum* L.)	*T. atroviride* and *Bacillus amyloliquefaciens*	*F. graminearum*	(Karuppiah et al., 2020)

Due to the multifaceted nature of agroecosystems, a complex network of interactions involving several actors is expected, including interactions between *Trichoderma* spp., phytopathogens, and other beneficial microbes, such as arbuscular mycorrhizal fungi (AMF) and growth-promoting rhizobacteria (GPR). The literature already reports these complex interactive networks that occur in the microbiota of the plant growth medium (Table 3). Studying and understanding such tripartite systems can reveal an essential cross-relation between interaction partners, which may not happen in bipartite interactions (Alfiky and Weisskopf, 2021).

The beneficial effects of *Trichoderma*-based biocontrol products can be measured from the increasing number of compounds available on the market, used in various crops (Asad, 2022). Revealing more about the complex biocontrol signaling network between *Trichoderma*, plants, and phytopathogens under conditions that are as close as possible to natural ones will allow scientists and industry to develop and advance molecular tools to improve the effectiveness and consistency of *Trichoderma* biocontrol strategies (Alfiky and Weisskopf, 2021).

Weed Control

Weeds are considered exotic invasive species that, as their reproduction is fast, compete for space and availability of soil nutrients with the crop (Schaffner et al., 2020). Controlling weeds has always been costly for three main reasons: time, yield, and cost. Based on yield estimates, the losses can be up to 35% (Triolet et al., 2020).

After World War II, the chemical industry developed molecules similar to plant hormones. The increasing specificity to control invasive plants made these compounds to be intensively consumed all over the globe. The most used method to contain invasive plants is the application of agrochemicals, which were developed to facilitate and optimize work in the agricultural sector. However, these agrochemicals cause serious environmental problems such as soil, water, and air contamination, negatively impacting human and animal health. In addition, some weeds resisted these products, making crop management harder. The scientific proof of these impacts resulted in movements in Europe, such as the establishment of a policy to reduce phytosanitary effects (Directive 2009/128/EC) so that greater adherence to the sustainable use of integrated pest management (IPM) could occur (Chauvel et al., 2012; Radhakrishnan et al., 2018; Triolet et al., 2020).

Sustainable alternatives to overcome the excessive use of agrochemicals have been explored, and biological herbicide agents are the flagship of research in this area. Fungi are generally good biological control agents because of some of their characteristics, such as their broad spectrum of action, easy maintenance, and adaptability to temperature, humidity, and the environment in which they develop, among others. In addition, under appropriate conditions, some fungi are capable of producing and secreting natural phytotoxic substances that function as possible bioherbicidal agents (Triolet et al., 2020). Thus, bioherbicides can be conceptualized as products containing living microorganisms and metabolites produced by them, which can control weeds in conventional agriculture following the IPM guidelines and in organic agriculture (Hoagland, 1990).

Regulations on plant protection products provide subsidies for biological products to enter the global market effectively. There are already some fungal-based natural products on the market, especially in the United States of America and Canada. Consolidated companies such as Bayer CropScience, BASF, and AgriScience are increasingly showing interest in research and product development in bioherbicides (Olson, 2015). A fact worth mentioning is that even if a product is of biological origin, it does not necessarily mean it is not toxic to the environment and animals. Therefore, investigating its spectrum of action is recommended (Triolet et al., 2020).

Colletotrichum, *Alternaria*, *Puccinia*, *Phoma*, and *Fusarium* are weed biocontrol's most studied fungal genera. However, in practice, their commercial development still needs to be improved, despite being successful candidates. In this scenario, the genus *Trichoderma* appears, which has a versatile genome, allowing the fungus to adapt to different regions of the world, including stressful environments and extreme conditions. Furthermore, *Trichoderma* spp. is famous for producing secondary metabolites, such as mycotoxins, enzymes, and organic compounds that present biocontrol potential (Al-Ani, 2019).

The methods used by *Trichoderma* to inhibit the growth of invasive plants and act as a bioherbicide are diverse, such as the (i) production of organic compounds, (ii) production of enzymes, (iii) production of other secondary metabolites, (iv) direct action on the plant and (v) induction of the plant's defense system. The effects that these mechanisms cause on weeds can be listed as (i) cytotoxicity producing phytotoxins, (ii) micronutrient deficiency, (iii) vascular wilt, (iv) leaf rust, and (v) leaf, stems, and roots rot and necrosis. The existence of other indirect interactions between secondary metabolites of *Trichoderma* and plants cannot be ruled out. Thus, the potential of this fungus and its metabolites to replace or reduce the use of conventional agrochemicals is clear (Al-Ani, 2019; Bejarano and Puopolo, 2020).

Applications of several *Trichoderma* species and their secondary metabolites demonstrated a reduction in the growth of weeds, which directly impacts crop yield. Table 4 compiles some studies that evaluated the use and effect of different species of *Trichoderma* on weeds.

Table 4. Use of *Trichoderma* in the control of weeds in crops

Trichoderma species	Target plant/weed	Observed effects/symptoms	Reference
T. koningiopsis	*Cucumis sativus* (model plant to evaluate herbicide effect)	Light depigmentation; Total depigmentation; Yellowing; Chlorosis; Necrosis.	(Ulrich et al., 2021)
Trichoderma spp. and *Prunus persica* L.	Soil application: evaluation of the occurrence of invasive plants and their relationship with Glicine max	Reduction in the frequency of weeds; Removal of wed biomass.	(Imran et al., 2020)
T. polysporum	*Avena fatua* L., *Polygonum lapathifolium* L., *Lepyrodiclis holosteoides*, *Chenopodium album* L., *Elsholtzia densa* and *Polygonum aviculare*	Yellowing; Dark brown lesions on leaf surface; Necrosis; Fresh mass inhibition; Plant death.	(Zhu et al., 2020)
T. harzianum, *T. pseudokoningi*, *T. reesei* and *T. viride*	*P. minor* and *R. dentatus*	Reduction in stalk length and fresh biomass; Formation of necrotic spots on the detached leaves.	(Javaid and Ali, 2011)
T. brevicompactum (CGMCC19618)	Monocot weeds: *Setaria viridis* L. Beauv. and *Echinochloa crusgalli* L. Beauv.	Reduction in the length of the root and upper part of the plants.	(Yin et al., 2020)
T. koningiopsis	Bidens pilosa and Euphorbia heterophylla	Anatomical transformations in the leaves of the plant; Necrosis.	(Camargo et al., 2020)
T. koningiopsis	*Bidens pilosa*, *Conyza bonariensis*, *Urochloa plantaginea*, and *Euphorbia heterophylla*	Leaf stress; Necrosis; Yellowing.	(Stefanski et al., 2020)
T. koningiopsis	*Euphorbia heterophylla*, *Bidens* sp., and *Urochloa plantaginea*	Contact damage; Hypersensitivity responses; Infection of leaf tissues; Leaf spot; Necrosis	(Reichert Júnior et al., 2019)

Trichoderma presents numerous possibilities that can be explored for bioherbicides. The diversity of symptoms and effects that species of this fungus cause on target plants suggests that an arsenal of metabolites and mechanisms is used to colonize its host. Therefore, a well-accepted hypothesis is that the fungus and its secondary metabolites can be used as biocontrol agents.

In most scenarios, *Trichoderma* uses more than one mechanism to act as a biocontrol agent. Directly, it can colonize host plants and consume available substrates, which could be the weed's leaf area (Bejarano and Puopolo, 2020). Indirectly, the compounds produced and released by the fungus may also cause an herbicidal effect on the affected weeds. An example of this is *T. virens*, which makes compounds such as Viridiol and the molecules (3H)-benzoxazolinone (BOA) and 2,4-dihydroxy-1,4-(2H)-benzoxazine-3-one (DIBOA) (Al-Ani and Mohammed, 2020; Keswani et al., 2013).

Another *Trichoderma* action that can be concomitant to direct colonization and that presents a herbicidal effect is the production and release of enzymes that catalyze several biochemical reactions in plants, which can help in the fight against weeds that cause damage to crops. Stefanski et al. (2020) quantified the enzymatic activity generated by the fermentation of *T. koningiopsis* and microalgae, demonstrating the presence of cellulases, lipases, amylases, and peroxidases. When the fungus penetrates the plant, it seeks to absorb nutrients, many of them trapped in molecules that are not directly usable by the fungus, so enzymes are tools that help the pathogen obtain these nutrients. Enzymes such as cellulases act on cellulose, a polysaccharide in all plant cells. Then, when this enzyme comes into contact with plant tissues, it breaks down the cellulose, separating it into smaller units and transforming cellulose into glucose (Gallon et al., 2009). Lipases, however, act by catalyzing reactions in lipids and producing fatty acids. That way, when made and released by *Trichoderma*, the lipases act on phospholipids, the main compound of plant cell walls, releasing these nutrients for the fungus to absorb (Pio et al., 2008).

Another essential substance present in plant tissues is starch, which is also not absorbed directly by the pathogen, so the tool used by the fungus to release the desired nutrients is the enzyme amylase, which is responsible for breaking down starch molecules to maltose and glucose that is absorbed quickly by the microorganism (Brito et al., 2015). Peroxidases are another class of enzymes produced by *Trichoderma* that act on various substrates such as plant phenolic compounds, proteins, peptides, and secondary plant metabolites, among others

(Carneiro et al., 2003). It is suggested that peroxidases act on the cell wall of plant tissues, causing damage and stress to the plants (Ulrich et al., 2021).

Different *Trichoderma* species also produce mycotoxins and trichothecenes, which are generally phytotoxic. Trichothecenes can cause necrosis, chlorosis, and mortality in plants. The phytotoxicity mechanism is believed to occur by blocking protein synthesis, which makes them potential herbicidal compounds (Yin et al., 2020).

Although these are natural compounds extracted or excreted by fungi of the genus *Trichoderma* to be used in the control of the weed, they, as well as their derivatives, must not be toxic to mammals. Considering that these compounds have high phytotoxicity and act as potential bioherbicides, they may come into contact with humans and animals and should not cause cellular damage in living beings. Therefore, it is recommended that when the herbicidal action of compounds such as trichothecenes is identified, cytotoxicity assessments be made in cell lines to understand mechanisms such as the average inhibitory concentration of the compound (Yin et al., 2020).

Plant Growth Promoter

In addition to the applications mentioned above, fungi of the genus *Trichoderma* have another function in agriculture: plant growth promoters. The capacity of *Trichoderma* to associate with plants and promote their growth has been cited in the literature for many years (Chang et al., 1986; Lindsey and Baker, 1967; Stewart and Hill, 2014). It is known that plants and microorganisms associate, mainly in the root region, which directly influences the ability of plants to obtain nutrients and, consequently, have better growth and development.

For a long time, the central hypothesis for the *Trichoderma* contribution to plant development was related to the ability to control and weaken phytopathogenic crop agents. However, this cannot be the only mechanism that promotes plant growth, as this effect was also observed in cultivation conditions without phytopathogenic microorganisms. Eventually, other tools were elucidated, such as the synthesis of phytohormones, solubilization of nutrients, production of vitamins, increase in root development, increase in absorption and translocation of nutrients, improvement in carbohydrate metabolism, photosynthesis and plant defense (Harman, 2006; Inbar et al., 1994; Stewart and Hill, 2014). However, it is worth noting that within the same species of *Trichoderma*, not all isolates can promote plant growth. Therefore, the interaction between the microorganism strain and the plant species must be considered (Tančić-Živanov et al., 2020; Tucci et al., 2011).

Benefits in plant growth after treatment with *Trichoderma* demonstrate a dependence on the plant species. For example, in one series of experiments, six strains of *Trichoderma* applied as a dry powder (obtained from submerged fermentation) consistently promoted the growth of lettuce seedlings. The strains increased the dry shoot weight by up to 26%. Some strains inhibited germination but increased shoots' fresh and dry weights (Ousley et al., 1994). Baker (1989) tested the same isolate on peas and radishes and found only small increases in growth. Likewise, when conidia suspensions of *Trichoderma* species were added to the soil, dry weight increased in pepper, cucumber, and tomato, but not with radish and bean plants (Chang et al., 1986).

The promoter/inhibitory effects exhibited by the same strain are possibly due to the differential production of secondary metabolites, some of which can be toxic to the plant. On the other hand, the differences may be related to the differential ability of the isolates to colonize plant roots (Stewart and Hill, 2014). Despite the inhibitory effect in some plants, several studies show the power of *Trichoderma* to promote growth in several species, as shown in Table 5.

Table 5. The growth-promoting effect of *Trichoderma* spp. on different plant species

Trichoderma species	Plant species	Effects	Reference
T. harzianum	*Myracrodruon urundeuva*	Increase in size, root dry mass, shoot dry mass, and root length	(Chagas Junior et al., 2022)
T. harzianum	*Lotus corniculatus* L.	Increase in total dry mass	(Machado et al., 2011)
Trichoderma spp. (several species)	*Vigna unguiculata* L. Walp.	Increase in plant height, root length, and total dry mass	(Chagas et al., 2016)
Trichoderma spp. (several species)	*P. vulgaris* L.	Increase in hypocotyl size and shoot dry mass	(Mayo-Prieto et al., 2020)
T. viride	*Fragaria x ananassa* Duch.	Increase in shoot dry mass and root dry mass	(Malusa et al., 2007)
T. harzianum	*T. aestivum* L.	Increase in the number of tillers	(Sharma et al., 2012)
T. asperellum	*Zea mays* L.	Increase in shoot and root length	(Tiru et al., 2021)
T. viride	*G. max* (L.) Merr.	Increase in shoot and root length	(John et al., 2010)

The type of inoculum and its formulation also influence the ability of *Trichoderma* to promote plant growth. Studies reported better plant development in several species when the inoculum was peat-based compared to conidial suspension (Baker et al., 1984). There are already commercial products based on *Trichoderma* for seed inoculation, usually carried out together with other microorganisms such as bacteria of the genera *Bradyrhizobium*, *Rhizobium*, and *Azospirillum*. The combination of *Trichoderma* isolates associated with bacteria such as *Bradyrhizobium* can promote better initial development of soybean plants, increasing the total dry mass of the root and shoot (Jacques et al., 2021).

Therefore, fungi of the genus *Trichoderma* are a promising alternative as a supplement for plant growth. They can be used in several plant species; however, care with the formulation and type of inoculation must be taken to obtain the best possible results.

Conclusion

The broad agricultural applications of fungi of the genus *Trichoderma* highlight it as a promising tool for the biological management of several pests and diseases that affect crops. From controlling diseases and weeds to promoting plant growth, this genus of fungi has enormous plasticity. It is present in several areas of agriculture, helping develop more sustainable management of agricultural ecosystems.

References

Adnan, M., Islam, W., Shabbir, A., Khan, K. A., Ghramh, H. A., Huang, Z., Chen, H. Y. H. and Lu, G. (2019). Plant defense against fungal pathogens by antagonistic fungi with *Trichoderma* in focus. *Microbial Pathogenesis,* 129:7–18. doi: 10.1016/j.micpath.2019.01.042.

Adsul, M., Sandhu, S. K., Singhania, R. R., Gupta, R., Puri, S. K. and Mathur, A. (2020). Designing a cellulolytic enzyme cocktail for the efficient and economical conversion of lignocellulosic biomass to biofuels. *Enzyme and Microbial Technology,* 133:109442. doi: 10.1016/j.enzmictec.2019.109442.

Al-Ani, L. K. T. (2019). Bioactive secondary metabolites of *Trichoderma* spp. for efficient management of phytopathogens. In *Secondary Metabolites of Plant Growth Promoting Rhizomicroorganisms: Discovery and Applications,* edited by H. B. Singh,

C. Keswani, M. S. Reddy, E. Sansinenea and C. García-Estrada, 125-143, Springer, Singapore. doi: 10.1007/978-981-13-5862-3_7.

Al-Ani, L. K. T. and Mohammed, A. M. (2020). Versatility of *Trichoderma* in plant disease management. In *Molecular Aspects of Plant Beneficial Microbes in Agriculture,* edited by V. Sharma, R. Salwan and L. K. T. Al-Ani, 159-168. Academic Press. doi: 10.1016/B978-0-12-818469-1.00013-4.

Alfiky A. and Weisskopf L. (2021). Deciphering *Trichoderma*–plant–pathogen interactions for better development of biocontrol applications. *Journal of Fungi,* 7(1):61. doi: 10.3390/jof7010061.

Ali, W. B., Chaduli, D., Navarro, D., Lechat, C., Turbé-Doan, A., Bertrand, E., Faulds, C. B., Sciara, G., Lesage-Meessen, L., Record, E. and Mechichi, T. (2020). Screening of five marine-derived fungal strains for their potential to produce oxidases with laccase activities suitable for biotechnological applications. *BMC Biotechnology,* 20(1):27. doi: 10.1186/s12896-020-00617-y.

Almeida, F., Rodrigues, M. L. and Coelho, C. (2019). The still underestimated problem of fungal diseases worldwide. *Frontiers in Microbiology,* 10:214. doi: 10.3389/fmicb. 2019.00214.

Amin, F., Razdan, V. K., Mohiddin, F. A., Bhat, K. A. and Banday, S. (2010). Potential of *Trichoderma* species as biocontrol agents of soil borne fungal propagules. *Journal of Phytology,* 2(10):38-41.

Aria, M. and Cuccurullo, C. (2017). Bibliometrix: an R-tool for comprehensive science mapping analysis. *Journal of Informetrics,* 11(4):959–975. doi: 10.1016/j.joi.2017. 08.007.

Asad, S. A. (2022). Mechanisms of action and biocontrol potential of *Trichoderma* against fungal plant diseases: a review. *Ecological Complexity,* 49:100978. doi: 10.1016/ j.ecocom.2021.100978.

Bai, B., Liu, C., Zhang, C., He, X., Wang, H., Peng, W. and Zheng, C. (2022). *Trichoderma* species from plant and soil: an excellent resource for biosynthesis of terpenoids with versatile bioactivities. *Journal of Advanced Research,* In Press. doi: 10.1016/ j.jare.2022.09.010.

Baiyee, B., Pornsuriya, C., Ito, S. and Sunpapao, A. (2019). *Trichoderma* spirale T76-1 displays biocontrol activity against leaf spot on lettuce (*Lactuca sativa* L.) caused by *Corynespora cassiicola* or *Curvularia aeria*. *Biological Control,* 129:195–200. doi: 10.1016/j.biocontrol.2018.10.018.

Baker, R. (1989). Improved *Trichoderma* spp. for promoting crop productivity. *Trends in Biotechnology,* 7(2):34–38. doi: 10.1016/0167-7799(89)90055-3.

Baker, R., Elad, Y. and Chet, I. (1984). The controlled experiment in the scientific method with special emphasis on biological control. *Phytopathology,* 74(9):1019. doi: 10.1094/Phyto-74-1019.

Baskaran, R. and Krishnan, C. (2020). Enhanced production of cellulase from a novel strain *Trichoderma gamsii* M501 through response surface methodology and its application in biomass saccharification. *Process Biochemistry,* 99:48–60. doi: 10.1016/j.procbio. 2020.08.006.

Bedine, M. A. B., Iacomi, B., Tchameni, S. N., Sameza, M. L. and Fekam, F. B. (2022). Harnessing the phosphate-solubilizing ability of *Trichoderma* strains to improve plant growth, phosphorus uptake and photosynthetic pigment contents in common bean (*Phaseolus vulgaris*). *Biocatalysis and Agricultural Biotechnology,* 45:102510. doi: 10.1016/j.bcab.2022.102510.

Bejarano, A. and Puopolo, G. (2020). Bioformulation of microbial biocontrol agents for a sustainable agriculture. In *How Research Can Stimulate the Development of Commercial Biological Control Against Plant Diseases, Progress in Biological Control,* edited by A. De Cal, P. Melgarejo and N. Magan, 275-293. Springer International Publishing. doi: 10.1007/978-3-030-53238-3_16.

Bellini, A., Gilardi, G., Idbella, M., Zotti, M., Pugliese, M., Bonanomi, G. and Gullino, M. L. (2023). *Trichoderma* enriched compost, BCAs and potassium phosphite control *Fusarium* wilt of lettuce without affecting soil microbiome at genus level. *Applied Soil Ecology,* 182:104678. doi: 10.1016/j.apsoil.2022.104678.

Bello, F., Montironi, I. D., Medina, M. B., Munitz, M. S., Ferreira, F. V., Williman, C., Vázquez, D., Cariddi, L. N. and Musumeci, M. A. (2022). Mycofumigation of postharvest blueberries with volatile compounds from *Trichoderma atroviride* IC-11 is a promising tool to control rots caused by *Botrytis cinerea. Food Microbiology,* 106:104040. doi: 10.1016/j.fm.2022.104040.

Benítez, T., Rincón, A. M., Limón, M. C. and Codón, A. C. (2004). Biocontrol mechanisms of *Trichoderma* strains. *International Microbiology,* 7:249–260.

Bilal, M. and Iqbal, H. M. N. (2020). Microbial peroxidases and their unique catalytic potentialities to degrade environmentally related pollutants. In *Microbial Technology for Health and Environment, Microorganisms for Sustainability,* edited by P. K. Arora, 1-24. Springer, Singapore. doi: 10.1007/978-981-15-2679-4_1.

Bischof, R. H., Ramoni, J. and Seiboth, B. (2016). Cellulases and beyond: the first 70 years of the enzyme producer *Trichoderma reesei. Microbial Cell Factories,* 15(1):106. doi: 10.1186/s12934-016-0507-6.

Bonatto, C., Scapini, T., Camargo, A. F., Alves Jr., S. L., Fongaro, G., de Oliveira, D. and Treichel, H. (2023). Microbiology of biofuels: cultivating the future. In *Relationship between Microbes and the Environment for Sustainable Ecosystem Services: Microbial Tools for Sustainable Ecosystem Services,* edited by J. Samuel, A. Kumar and J. Singh, 342. Elsevier. doi: 10.1016/B978-0-323-89936-9.00005-9.

Bordin, E. R., Camargo, A. F., Stefanski, F. S., Scapini, T., Bonatto, C., Zanivan, J., Preczeski, K., Modkovski, T. A., Reichert Júnior, F. W., Mossi, A. J., Fongaro, G., Ramsdorf, W. A. and Treichel, H. (2021). Current production of bioherbicides: mechanisms of action and technical and scientific challenges to improve food and environmental security. *Biocatalysis and Biotransformation,* 39(5):346–359. doi: 10.1080/10242422.2020.1833864.

Brito, V. H. dos S., da Silva, E. C. and Cereda, M. P. (2015). Digestibilidade do amido *in vitro* e valor calórico dos grupos de farinhas de mandioca brasileiras [*In vitro* starch digestibility and caloric value of Brazilian cassava flour groups]. *Braz J Food Technol,* 18:185–191. doi: 10.1590/1981-6723.2714.

Camargo, A. F., Stefanski, F. S., Scapini, T., Weirich, S. N., Ulkovski, C., Carezia, C., Bordin, E. R., Rossetto, V., Reichert Júnior, F. W., Galon, L., Fongaro, G., Mossi, A. J. and Treichel, H. (2019). Resistant weeds were controlled by the combined use of herbicides and bioherbicides. *Environmental Quality Management,* 29(1):37–42. doi: 10.1002/tqem.21643.

Camargo, A. F., Venturin, B., Bordin, E. R., Scapini, T., Stefanski, F. S., Klanovicz, N., Dalastra, C., Kubeneck, S., Preczeski, K. P., Rossetto, V., Weirich, S., Carezia, C., Ulkovski, C., Reichert Júnior, F. W., Müller, C., Fongaro, G., Mossi, A. J. and Treichel, H. (2020). A low-genotoxicity bioherbicide obtained from *Trichoderma koningiopsis* fermentation in a stirred-tank bioreactor. *Industrial Biotechnology,* 16(3):176–181. doi: 10.1089/ind.2019.0024.

Carneiro, C. E. A., Rolim, H. M. V. and Fernandes, K. F. (2003). Estudo das atividades de peroxidades e polifenoloxidase de guariroba (*Syagrus oleracea* Becc) sob a ação de diferentes inibidores [Study of peroxidase and polyphenoloxidase activities of guariroba (*Syagrus oleracea* Becc) under the action of different inhibitors]. *Acta Sci Biol Sci,* 25(1):189–193. doi: 10.4025/actascibiolsci. v25i1.2108.

Carrero-Carrón, I., Trapero-Casas, J. L., Olivares-García, C., Monte, E., Hermosa, R. and Jiménez-Díaz, R. M. (2016). *Trichoderma asperellum* is effective for biocontrol of *Verticillium* wilt in olive caused by the defoliating pathotype of *Verticillium dahliae. Crop Protection,* 88:45–52. doi: 10.1016/j.cropro.2016.05.009.

Cavalcante, B. D. M., Scapini, T., Camargo, A. F., Ulrich, A., Bonatto, C., Dalastra, C., Mossi, A. J., Fongaro, G., Di Piero, R. M. and Treichel, H. (2021). Orange peels and shrimp shell used in a fermentation process to produce an aqueous extract with bioherbicide potential to weed control. *Biocatalysis and Agricultural Biotechnology,* 32:101947. doi: 10.1016/j.bcab.2021.101947.

Chagas Junior, A. F., Dias, P. C., dos Santos, G. R., Ribeiro, A. S. N., de Sousa, K. Â. O. and Chagas, L. F. B. (2022). *Trichoderma* as a growth promoter in *Astronium urundeuva* (M. Allemão). *Engl. Scientia Plena,* 18(5). doi: 10.14808/sci.plena.2022.056201.

Chagas, L. F. B., De Castro, H. G., Colonia, B. S. O., De Carvalho Filho, M. R., Miller, L. D. O. and Chagas, A. F. J. (2016). Efficiency of *Trichoderma* spp. as a growth promoter of cowpea (*Vigna unguiculata*) and analysis of phosphate solubilization and indole acetic acid synthesis. *Braz J Bot,* 39(2):437–445. doi: 10.1007/s40415-015-0247-6.

Chang, Y.-C., Chang, Y.-C., Baker, R., Kleifeld, O. and Chet, I. (1986). Increased growth of plants in the presence of the biological control agent *Trichoderma harzianum. Plant Dis,* 70(2):145. doi: 10.1094/PD-70-145.

Chauvel, B., Guillemin, J.-P., Gasquez, J. and Gauvrit, C. (2012). History of chemical weeding from 1944 to 2011 in France: changes and evolution of herbicide molecules. *Crop Protection,* 42:320–326. doi: 10.1016/j.cropro.2012.07.011.

Chen, S., Li, H., Chen, Y., Li, S., Xu, J., Guo, H., Liu, Z., Zhu, S., Liu, H. and Zhang, W. (2019). Three new diterpenes and two new sesquiterpenoids from the endophytic fungus *Trichoderma koningiopsis* A729. *Bioorganic Chemistry*, 86:368–374. doi: 10.1016/j.bioorg.2019.02.005.

Chovanec, P., Kaliňák, M., Liptaj, T., Pronayová, N., Jakubík, T., Hudecová, D. and Varečka, Ľ. (2005). Study of *Trichoderma viride* metabolism under conditions of the restriction of oxidative processes. *Can J Microbiol,* 51(10):853–862. doi: 10.1139/w05-075.

Correa, L. O., Bezerra, A. F. M., Honorato, L. R. S., Cortez, A. C. A., Souza, J. V. B. and Souza, E. S. (2021). Amazonian soil fungi are efficient degraders of glyphosate herbicide; novel isolates of *Penicillium*, *Aspergillus*, and *Trichoderma*. *Braz J Biol,* 83. doi: 10.1590/1519-6984.242830.

da Costa, A. C., de Miranda, R. F., Costa, F. A. and Ulhoa, C. J. (2021). Potential of *Trichoderma piluliferum* as a biocontrol agent of *Colletotrichum musae* in banana fruits. *Biocatalysis and Agricultural Biotechnology,* 34:102028. doi: 10.1016/j.bcab.2021.102028.

da Silveira, A. A., Andrade, J. S. P., Guissoni, A. C. P., da Costa, A. C., de Carvalho e Silva, A., da Silva, H. G., Brito, P., de Souza, G. R. L. and Fernandes, K. F. (2021). Larvicidal potential of cell wall degrading enzymes from *Trichoderma asperellum* against *Aedes aegypti* (Diptera: Culicidae). *Biotechnology Progress,* 37(5):e3182. doi: 10.1002/btpr.3182.

de Oliveira, M. R., Katak, R. de M., da Silva, G. F., Marinotti, O., Terenius, O., Tadei, W. P., de Souza, A. D. L. and de Souza, A. Q. L. (2021). Extracts of amazonian fungi with larvicidal activities against *Aedes aegypti*. *Frontiers in Microbiology,* 10;12:743246. doi: 10.3389/fmicb.2021.743246.

De Paula, N. M., da Silva, K., Brugnari, T., Haminiuk, C. W. I. and Maciel, G. M. (2022). Biotechnological potential of fungi from a mangrove ecosystem: enzymes, salt tolerance and decolorization of a real textile effluent. *Microbiological Research,* 254:126899. doi: 10.1016/j.micres.2021.126899.

Dell'Anno, F., Rastelli, E., Buschi, E., Barone, G., Beolchini, F. and Dell'Anno, A. (2022). Fungi can be more effective than bacteria for the bioremediation of marine sediments highly contaminated with heavy metals. *Microorganisms,* 10(5):993. doi: 10.3390/microorganisms10050993.

Doehlemann, G., Ökmen, B., Zhu, W. and Sharon, A. (2017). Plant pathogenic fungi. *Microbiology Spectrum,* 5(1):5.1.14. doi: 10.1128/microbiolspec.FUNK-0023-2016.

Druzhinina, I. S., Seidl-Seiboth, V., Herrera-Estrella, A., Horwitz, B. A., Kenerley, C. M., Monte, E., Mukherjee, P. K., Zeilinger, S., Grigoriev, I. V. and Kubicek, C. P. (2011). *Trichoderma*: the genomics of opportunistic success. *Nat Rev Microbiol,* 9(10):749–759. doi: 10.1038/nrmicro2637.

Dutta, P., Deb, L. and Pandey, A. K. (2022). *Trichoderma*- from lab bench to field application: looking back over 50 years. *Frontiers in Agronomy,* 4:932839. doi: doi.org/10.3389/fagro.2022.932839.

El-Komy, M. H., Al-Qahtani, R. M., Ibrahim, Y. E., Almasrahi, A. A., Al-Saleh, M. A. (2022). Soil application of *Trichoderma asperellum* strains significantly improves *Fusarium* root and stem rot disease management and promotes growth in cucumbers in semi-arid regions. *Eur J Plant Pathol,* 162(3):637–653. doi: 10.1007/s10658-021-02427-0.

Errasquín, E. L. and Vázquez, C. (2003). Tolerance and uptake of heavy metals by *Trichoderma atroviride* isolated from sludge. *Chemosphere,* 50(1):137–143. doi: 10.1016/S0045-6535(02)00485-X.

Escudero-Leyva, E., Alfaro-Vargas, P., Muñoz-Arrieta, R., Charpentier-Alfaro, C., Granados-Montero, M. del M., Valverde-Madrigal, K. S., Pérez-Villanueva, M., Méndez-Rivera, M., Rodríguez-Rodríguez, C. E., Chaverri, P. and Mora-Villalobos,

J. A. (2022). Tolerance and biological removal of fungicides by *Trichoderma* species isolated from the endosphere of wild Rubiaceae Plants. *Frontiers in Agronomy,* 3:772170. doi: 10.3389/fagro.2021.772170.

Fang, H., Zhao, C., Li, C., Song, Y., Yu, L., Song, X., Wu, J. and Yang, L. (2023). Direct consolidated bioprocessing for d-glucaric acid production from lignocellulose under subcritical water pretreatment. *Chemical Engineering Journal,* 454:140339. doi: 10.1016/j.cej.2022.140339.

Felix, C. R., Noronha, E. F. and Miller, R. N. G. (2014). *Trichoderma*: a dual function fungi and their use in the wine and beer industries. In *Biotechnology and Biology of Trichoderma,* edited by V. K. Gupta, M. Schmoll, A. Herrera-Estrella, R. S. Upadhyay, I. Druzhinina and M. G. Tuohy, 345-349. Elsevier, Amsterdam. doi: 10.1016/B978-0-444-59576-8.00025-4.

Freeman, S., Minz, D., Kolesnik, I., Barbul, O., Zveibil, A., Maymon, M., Nitzani, Y., Kirshner, B., Rav-David, D., Bilu, A., Dag, A., Shafir, S. and Elad, Y. (2004). *Trichoderma* biocontrol of *Colletotrichum acutatum* and *Botrytis cinerea* and survival in strawberry. *European Journal of Plant Pathology,* 110(4):361–370. doi: 10.1023/B:EJPP.0000021057.93305.d9.

Gallon, C. Z., Broetto, S. G., Silva, D. M. (2009). Atividade da celulase e β-galactosidase no estudo da firmeza da polpa de mamões 'golden' e 'gran golden' [Cellulase and β-galactosidase activity in the study of 'golden' and 'gran golden' papaya pulp firmness]. *Rev Bras Frutic,* 31:1178–1183. doi: 10.1590/S0100-29452009000400035.

Geng, L., Fu, Y., Peng, X., Yang, Z., Zhang, M., Song, Z., Guo, N., Chen, S., Chen, J., Bai, B., Liu, A. and Ahammed, G. J. (2022). Biocontrol potential of *Trichoderma harzianum* against *Botrytis cinerea* in tomato plants. *Biological Control,* 174:105019. doi: 10.1016/j.biocontrol.2022.105019.

Ghorbanpour, M., Omidvari, M., Abbaszadeh-Dahaji, P., Omidvar, R. and Kariman, K. (2018). Mechanisms underlying the protective effects of beneficial fungi against plant diseases. *Biological Control,* 117:147–157. doi: 10.1016/j.biocontrol.2017.11.006.

Gilardi, G., Pugliese, M., Gullino, M. L. and Garibaldi, A. (2019). Nursery treatments with resistant inducers, soil amendments and biocontrol agents for the management of the *Fusarium wilt* of lettuce under glasshouse and field conditions. *Journal of Phytopathology,* 167(2): 98–110. doi: 10.1111/jph.12778.

Gveroska, B. and Ziberoski, J. (2012). *Trichoderma harzianum* as a biocontrol agent against *Alternaria alternata* on tobacco. *Applied Technologies and Innovations,* 7(2):67–77. doi: 10.15208/ati.2012.9.

Hanada, R. E., Pomella, A. W. V., Soberanis, W., Loguercio, L. L. and Pereira, J. O. (2009). Biocontrol potential of *Trichoderma martiale* against the black-pod disease (*Phytophthora palmivora*) of cacao. *Biological Control,* 50(2):143–149. doi: 10.1016/j.biocontrol.2009.04.005.

Hang, X., Meng, L., Ou, Y., Shao, C., Xiong, W., Zhang, N., Liu, H., Li, R., Shen, Q. and Kowalchuk, G. A. (2022). *Trichoderma*-amended biofertilizer stimulates soil resident *Aspergillus* population for joint plant growth promotion. *Biofilms Microbiomes,* 8(1):1–7. doi: 10.1038/s41522-022-00321-z.

Harman, G. E. (2006). Overview of mechanisms and uses of *Trichoderma* spp. *Phytopathology,* 96(2):190–194. doi: 10.1094/PHYTO-96-0190.

Harris, A. R. (1999). Biocontrol of *Rhizoctonia solani* and *Pythium ultimum* on *Capsicum* by *Trichoderma koningii* in potting medium. *Microbiological Research,* 154(2):131–135. doi: 10.1016/S0944-5013(99)80005-6.

Hermosa, R., Viterbo, A., Chet, I. and Monte, E. (2012). Plant-beneficial effects of *Trichoderma* and of its genes. *Microbiology,* 158(1):17–25. doi: 10.1099/mic.0.052274-0.

Hoagland, R. E. (1990). Microbes and microbial products as herbicides. In Microbes and Microbial Products as Herbicides, 2-52. *ACS Symposium Series, American Chemical Society.* doi: 10.1021/bk-1990-0439.ch001.

Imran, Amanullah, Arif, M., Shah, Z. and Bari, A. (2020). Soil application of *Trichoderma* and peach (*Prunus persica* L.) residues possesses biocontrol potential for weeds and enhances growth and profitability of soybean (*Glycine max*). *Sarhad Journal of Agriculture,* 36:10–20. doi: 10.17582/journal.sja/2020/36.1.10.20.

Inbar, J., Abramsky, M., Cohen, D. and Chet I. (1994). Plant growth enhancement and disease control by *Trichoderma harzianum* in vegetable seedlings grown under commercial conditions. *Eur J Plant Pathol,* 100(5):337–346. doi: 10.1007/BF01876444.

Jacques, A. P., Martins, M. P., Santana, A. B., Chiqueto, G. da S. G., Neto, J. F., Ulhoa, C. J. and Costa, F. A. (2021). Isolados nativos de *Trichoderma* spp. como promotor de crescimento na fase inicial da cultura da soja [Native isolates of *Trichoderma* spp. as a growth promoter in the initial phase of soybean culture]. *Brazilian Journal of Development,* 7(11):108150–108166. doi: 10.34117/bjdv7n11-441.

Jain, A., Sarsaiya, S., Wu, Q., Lu, Y. and Shi, J. (2019). A review of plant leaf fungal diseases and its environment speciation. *Bioengineered,* 10(1):409–424. doi: 10.1080/21655979.2019.1649520.

Jaroszuk-Ściseł, J., Tyśkiewicz, R., Nowak, A., Ozimek, E., Majewska, M., Hanaka, A., Tyśkiewicz, K., Pawlik, A. and Janusz, G. (2019). Phytohormones (auxin, gibberellin) and ACC deaminase *in vitro* synthesized by the mycoparasitic *Trichoderma* DEMTkZ3A0 strain and changes in the level of auxin and plant resistance markers in wheat seedlings inoculated with this strain conidia. *International Journal of Molecular Sciences,* 20(19):4923. doi: 10.3390/ijms20194923.

Javaid, A. and Ali, S. (2011). Herbicidal activity of culture filtrates of *Trichoderma* spp. against two problematic weeds of wheat. *Natural Product Research,* 25(7):730–740. doi: 10.1080/14786419.2010.528757.

Jiang, Y., Wang, J.-L., Chen, J., Mao, L.-J., Feng, X.-X., Zhang, C.-L. and Lin, F.-C. (2016). *Trichoderma* biodiversity of agricultural fields in east China reveals a gradient distribution of species. *PLoS ONE,* 11(8):e0160613. doi: 10.1371/journal.pone.0160613.

John, R. P., Tyagi, R. D., Prévost, D., Brar, S. K., Pouleur, S. and Surampalli, R. Y. (2010). Mycoparasitic *Trichoderma viride* as a biocontrol agent against *Fusarium oxysporum* f. sp. adzuki and *Pythium arrhenomanes* and as a growth promoter of soybean. *Crop Protection,* 29(12):1452–1459. doi: 10.1016/j.cropro.2010.08.004.

Karuppiah, V., Li, Y., Sun, J., Vallikkannu, M. and Chen, J. (2020). Vel1 regulates the growth of *Trichoderma atroviride* during co-cultivation with *Bacillus amyloliquefaciens* and is essential for wheat root rot control. *Biological Control,* 151:104374. doi: 10.1016/j.biocontrol.2020.104374.

Kasonga, T. K., Coetzee, M. A. A., Kamika, I. and Momba, M. N. B. (2022). Assessing a co-culture fungal granule ability to remove pharmaceuticals in a sequencing batch reactor. *Environmental Technology,* 43(11):1684–1699. doi: 10.1080/09593330.2020.1847204.

Katayama, T., Miyagawa, K., Kodama, T. and Oikawa, S. (2001). Trichorzin HA V, a member of the peptaibol family, stimulates intracellular cAMP formation in cells expressing the calcitonin receptor. *Biological and Pharmaceutical Bulletin,* 24(12):1420–1422. doi: 10.1248/bpb.24.1420.

Keswani, C., Singh, S. P. and Singh, H. B. (2013). A superstar in biocontrol enterprise: *Trichoderma* spp. *Biotech Today,* 3(2):27–30. doi: 10.5958/2322-0996.2014.00005.2.

Kexiang, G., Xiaoguang, L., Yonghong, L., Tianbo, Z. and Shuliang, W. (2002). Potential of *Trichoderma harzianum* and *T. atroviride* to control *Botryosphaeria berengeriana* f. sp. piricola, the cause of apple ring rot. *Journal of Phytopathology,* 150(4–5):271–276. doi: 10.1046/j.1439-0434.2002.00754.x.

Khatso, K. and Ao, N. T. (2013). Biocontrol of rhizome rot disease of ginger (*Zingiber officinale* Rosc.). *International Journal of Bio-resource and Stress Management,* 4(2)special:317-321.

Kithan, C. and L, D. (2014). *In vitro* evaluation of botanicals, bio-agents and fungicides against leaf blight of *Etlingera linguiformis* caused by *Curvularia lunata* Var. Aeria. *Journal of Plant Pathology & Microbiology,* 5(3):1–6. doi: 10.4172/2157-7471.1000232.

Klanovicz, N., Stefanski, F. S., Camargo, A. F., Michelon, W., Treichel, H. and Teixeira, A. C. S. C. (2022). Complete wastewater discoloration by a novel peroxidase source with promising bioxidative properties. *Journal of Chemical Technology & Biotechnology,* 97(9):2613–2625. doi: 10.1002/jctb.7134.

Klaram, R., Jantasorn, A. and Dethoup, T. (2022). Efficacy of marine antagonist, *Trichoderma* spp. as halo-tolerant biofungicide in controlling rice diseases and yield improvement. *Biological Control,* 172:104985. doi: 10.1016/j.biocontrol.2022.104985.

Kullnig, C. M., Krupica, T., Woo, S. L., Mach, R. L., Rey, M., Benítez, T., Lorito, M. and Kubicek, C. P. (2001). Confusion abounds over identities of *Trichoderma* biocontrol isolates. *Mycological Research,* 105(7):770–772. doi: 10.1017/S0953756201229967.

Latha, P., Anand, T., Prakasam, V., Jonathan, E. I., Paramathma, M. and Samiyappan, R. (2011). Combining *Pseudomonas, Bacillus* and *Trichoderma* strains with organic amendments and micronutrient to enhance suppression of collar and root rot disease in physic nut. *Applied Soil Ecology,* 49:215–223. doi: 10.1016/j.apsoil.2011.05.003.

Li, M., Ma, G., Lian, H., Su, X., Tian, Y., Huang, W., Mei, J. and Jiang, X. (2019). The effects of *Trichoderma* on preventing cucumber *Fusarium wilt* and regulating cucumber physiology. *Journal of Integrative Agriculture,* 18(3):607–617. doi: 10.1016/S2095-3119(18)62057-X.

Lindsey, D. L. and Baker, R. (1967). Effect of certain fungi on dwarf tomatoes grown under gnotobiotic conditions. *Phytopathology,* 57:111262–1263.

Machado, R. G., Hahn, L., Almeida, D., Moraes, T., Camargo, A. de O. and Reartes, D. S. (2011). Promoção de crescimento de *Lotus corniculatus* L. e *Avena strigosa* Schreb pela inoculação conjunta de *Trichoderma harzianum* e rizóbio [Growth promotion of *Lotus corniculatus* L. and *Avena strigosa* Schreb by joint inoculation of *Trichoderma harzianum* and rhizobium]. *Ciência E Natura,* 33(2):111-126. doi: 10.5902/2179 460X9365.

Malmir, N., Zamani, M., Motallebi, M., Fard, N. A. and Mekuto, L. (2022). Cyanide biodegradation by *Trichoderma harzianum* and cyanide hydratase network analysis. *Molecules,* 27(10):3336. doi: 10.3390/molecules27103336.

Malusa, E., Sas-Paszt, L., Popinska, W. and Zurawicz, E. (2007). The effect of a substrate containing arbuscular mycorrhizal fungi and rhizosphere microorganisms (*Trichoderma*, *Bacillus*, *Pseudomonas* and *Streptomyces*) and foliar fertilization on growth response and rhizosphere pH of three strawberry cultivars. *International Journal of Fruit Science,* 6(4):25–41. doi: 10.1300/J492v06n04_04.

Mao, T., Chen, X., Ding, H., Chen, X. and Jiang X. (2020). Pepper growth promotion and *Fusarium wilt* biocontrol by *Trichoderma hamatum* MHT1134. *Biocontrol Science and Technology,* 30(11):1228–1243. doi: 10.1080/09583157.2020.1803212.

Martínez-Medina, A., Roldán, A., Albacete, A. and Pascual, J. A. (2011). The interaction with arbuscular mycorrhizal fungi or *Trichoderma harzianum* alters the shoot hormonal profile in melon plants. *Phytochemistry,* 72(2):223–229. doi: 10.1016/ j.phytochem.2010.11.008.

Mayo-Prieto, S., Campelo, M. P., Lorenzana, A., Rodríguez-González, A., Reinoso, B., Gutiérrez, S. and Casquero, P. A. (2020). Antifungal activity and bean growth promotion of *Trichoderma* strains isolated from seed vs soil. *Eur J Plant Pathol,* 158(4):817–828. doi: 10.1007/s10658-020-02069-8.

Mishra, D. S., Gupta, A., Prajapati, C. R. and Singh, U. (2011). Combination of fungal and bacterial antagonists for management of root and stem rot disease of soybean. *Pakistan Journal of Botany,* 43:2569–74.

Mostafa, A. A.-F., Yassin, M. T., Dawoud, T. M., Al-Otibi, F. O. and Sayed, S. R. M. (2022). Mycodegradation of diazinon pesticide utilizing fungal strains isolated from polluted soil. *Environmental Research,* 212:113421. doi: 10.1016/j.envres.2022. 113421.

Mwangi, M. W., Monda, E. O., Okoth, S. A. and Jefwa, J. M. (2011). Inoculation of tomato seedlings with *Trichoderma harzianum* and arbuscular mycorrhizal fungi and their effect on growth and control of wilt in tomato seedlings. *Braz J Microbiol,* 42(2):508–513. doi: 10.1590/S1517-83822011000200015.

Naglot, A., Goswami, S., Rahman, I., Shrimali, D. D., Yadav, K. K., Gupta, V. K., Rabha, A. J., Gogoi, H. K. and Veer, V. (2015). Antagonistic potential of native *Trichoderma viride* strain against potent tea fungal pathogens in north east India. *Plant Pathol J,* 31(3):278–289. doi: 10.5423/PPJ.OA.01.2015.0004.

Olowe, O. M., Nicola, L., Asemoloye, M. D., Akanmu, A. O. and Babalola, O. O. (2022). *Trichoderma*: potential bio-resource for the management of tomato root rot diseases in Africa. *Microbiological Research,* 257:126978. doi: 10.1016/j.micres.2022. 126978.

Olson, S. (2015). An analysis of the biopesticide market now and where it is going. *Outlooks on Pest Management,* 26(5):203–206. doi: 10.1564/v26_oct_04.

Ousley, M. A., Lynch, J. M. and Whipps, J. M. (1994). Potential of *Trichoderma* spp. as consistent plant growth stimulators. *Biol Fertil Soils,* 17(2):85–90. doi: 10.1007/BF00337738.

Pandey, S. and Pundhir, V. S. (2013). Mycoparasitism of potato black scurf pathogen (*Rhizoctonia solani* Kuhn) by biological control agents to sustain production. *Indian Journal of Horticulture,* 70(1):71–75.

Paymaneh, Z., Sarcheshmehpour, M., Mohammadi, H. and Hesni, M. A. (2023). Vermicompost and/or compost and arbuscular mycorrhizal fungi are conducive to improving the growth of pistachio seedlings to drought stress. *Applied Soil Ecology,* 182:104717. doi: 10.1016/j.apsoil.2022.104717.

Perera, D. S., Tharaka, W. G. H., Amarasinghe, D. and Wickramarachchi, S. R. (2023). Extracellular extracts of antagonistic fungi, *Trichoderma longibrachiatum* and *Trichoderma viride*, as larvicides against dengue vectors, *Aedes aegypti* and *Aedes albopictus*. *Acta Tropica,* 238:106747. doi: 10.1016/j.actatropica.2022.106747.

Pio, T. F., Fraga, L. P., Macedo, G. A. and Kamimura, E. S. (2008). Cutinases fúngicas: propriedades e aplicações industriais [Fungal cutinases: properties and industrial applications]. *Química Nova,* 31:2118–2123. doi: 10.1590/S0100-40422008000800036.

Podder, D. and Ghosh, S. K. (2019). A new application of *Trichoderma asperellum* as an anopheline larvicide for eco friendly management in medical science. *Sci Rep,* 9(1):1108. doi: 10.1038/s41598-018-37108-2.

Poveda, J. (2021). Biological control of *Fusarium oxysporum* f. sp. ciceri and *Ascochyta rabiei* infecting protected geographical indication Fuentesaúco-Chickpea by *Trichoderma* species. *Eur J Plant Pathol,* 160(4):825–840. doi: 10.1007/s10658-021-02286-9.

Radhakrishnan, R., Alqarawi, A. A. and AbdAllah, E. F. (2018). Bioherbicides: current knowledge on weed control mechanism. *Ecotoxicology and Environmental Safety,* 158:131–138. doi: 10.1016/j.ecoenv.2018.04.018.

Ramezani, H. (2008). Biological control of root-rot of eggplant caused by *Macrophomina phaseolina*. *America-Eurasian Journal of Agriculture and Environmental Sciences,* 4(2):218-220.

Rees, H. J., Drakulic, J., Cromey, M. G., Bailey, A. M. and Foster, G. D. (2022). Endophytic *Trichoderma* spp. can protect strawberry and privet plants from infection by the fungus *Armillaria mellea*. *PLOS ONE,* 17(8):e0271622. doi: 10.1371/journal.pone.0271622.

Reichert Júnior, F. W., Scariot, M. A., Forte, C. T., Pandolfi, L., Dil, J. M., Weirich, S., Carezia, C., Mulinari, J., Mazutti, M. A., Fongaro, G., Galon, L., Treichel, H. and Mossi, A. J. (2019). New perspectives for weeds control using autochthonous fungi with selective bioherbicide potential. *Heliyon,* 5(5):e01676. doi: 10.1016/j.heliyon.2019.e01676.

Saini, S., Kumar, A., Singhal, B., Kuhad, R. C., Sharma, K. K. (2022). Fungal oxidoreductases and CAZymes effectively degrade lignocellulosic component of switchgrass for bioethanol production. *Fuel,* 328:125341. doi: 10.1016/j.fuel.2022.125341.

Samuels, G. J. (1996). *Trichoderma*: a review of biology and systematics of the genus. *Mycological Research,* 100(8):923–935. doi: 10.1016/S0953-7562(96)80043-8.

Sandle, T. (2014). *Trichoderma.* In *Encyclopedia of Food Microbiology* edited by C. A. Batt and M. L. Tortorello, 644-646. Academic Press, Oxford. doi: 10.1016/B978-0-12-384730-0.00337-2.

Sanguine, I. S., Cavalheiro, G. F., Garcia, N. F. L., dos Santos, M. V., Gandra, J. R., de Goes, R. H. de T. e B., da Paz, M. F., Fonseca, G. G. and Leite R. S. R. (2022). Xylanases of *Trichoderma koningii* and *Trichoderma pseudokoningii*: production, characterization and application as additives in the digestibility of forage for cattle. *Biocatalysis and Agricultural Biotechnology,* 44:102482. doi: 10.1016/j.bcab.2022.102482.

Saravanakumar, K., Yu, C., Dou, K., Wang, M., Li, Y. and Chen, J. (2016). Synergistic effect of *Trichoderma*-derived antifungal metabolites and cell wall degrading enzymes on enhanced biocontrol of *Fusarium oxysporum* f. sp. cucumerinum. *Biological Control*, 94:37–46. doi: 10.1016/j.biocontrol.2015.12.001.

Saravanan, A., Kumar, P. S., Vo, D.-V .N., Jeevanantham, S., Karishma, S. and Yaashikaa, P. R. (2021). A review on catalytic-enzyme degradation of toxic environmental pollutants: microbial enzymes. *Journal of Hazardous Materials,* 419:126451. doi: 10.1016/j.jhazmat.2021.126451.

Sarhan, E. A. D., Abd-Elsyed, M. H. F. and Ebrahiem, A. M. Y. (2020). Biological control of cucumber powdery mildew (*Podosphaera xanthii*) (Castagne) under greenhouse conditions. *Egyptian Journal of Biological Pest Control,* 30(1):65. doi: 10.1186/s41938-020-00267-4.

Schaffner, U., Steinbach, S., Sun, Y., Skjøth, C. A., de Weger, L. A., Lommen, S. T., Augustinus, B. A., Bonini, M., Karrer, G., Šikoparija, B., Thibaudon, M. and Müller-Schärer, H. (2020). Biological weed control to relieve millions from *Ambrosia* allergies in Europe. *Nat Commun,* 11(1):1745. doi: 10.1038/s41467-020-15586-1.

Sebumpan, R., Guiritan, K. R., Suan, M., Abapo, C. J., Bhat, A. H., Machado, R. A. R., Nimkingrat, P. and Sumaya, N. H. (2022). Morphological and molecular identification of *Trichoderma asperellum* isolated from a dragon fruit farm in the southern Philippines and its pathogenicity against the larvae of the super worm, *Zophobas morio* (Fabricius, 1776) (*Coleoptera*: Tenebrionidae). *Egyptian Journal of Biological Pest Control,* 32(1):47. doi: 10.1186/s41938-022-00548-0.

Sennoi, R., Singkham, N., Jogloy, S., Boonlue, S., Saksirirat, W., Kesmala, T. and Patanothai, A. (2013). Biological control of southern stem rot caused by *Sclerotium rolfsii* using *Trichoderma harzianum* and arbuscular mycorrhizal fungi on Jerusalem artichoke (*Helianthus tuberosus* L.). *Crop Protection,* 54:148–153. doi: 10.1016/j.cropro.2013.08.011.

Shakeri, J. and Foster, H. A. (2007). Proteolytic activity and antibiotic production by *Trichoderma harzianum* in relation to pathogenicity to insects. *Enzyme and Microbial Technology,* 40(4):961–968. doi: 10.1016/j.enzmictec.2006.07.041.

Shankar, A., Saini, S. and Sharma, K. K. (2022). Fungal-integrated second-generation lignocellulosic biorefinery: utilization of agricultural biomass for co-production of lignocellulolytic enzymes, mushroom, fungal polysaccharides, and bioethanol. *Biomass Conv Bioref.* doi: 10.1007/s13399-022-02969-1.

Sharma, A., Salwan, R., Kaur, R., Sharma, R. and Sharma, V. (2022). Characterization and evaluation of bioformulation from antagonistic and flower inducing *Trichoderma asperellum* isolate UCRD5. *Biocatalysis and Agricultural Biotechnology,* 43:102437. doi: 10.1016/j.bcab.2022.102437.

Sharma, P., Patel, A., Saini, M. and Deep, S. (2012). Field demonstration of *Trichoderma harzianum* as a plant growth promoter in wheat (*Triticum aestivum* L). *Journal of Agricultural Science,* 4:65–73. doi: 10.5539/jas.v4n8p65.

Silva, H. F., Santos, A. M. G., do Amaral, A. C. T., Bezerra, J. L. and Luz, E. D. M. N. (2020). Bioprospection of *Trichoderma* spp. originating from a Cerrado-Caatinga ecotone on *Colletotrichum truncatum*, in soybean. *Revista Brasileira de Ciências Agrárias,* 15(1):1–7. doi: 10.5039/agraria.v15i1a7680.

Silva, R. N., Steindorff, A. S. and Monteiro, V. N. (2014). Metabolic diversity of *Trichoderma*. In *Biotechnology and Biology of Trichoderma*, edited by V. K. Gupta, M. Schmoll, A. Herrera-Estrella, R. S. Upadhyay, I. Druzhinina and M. G. Tuohy, 363-376. Elsevier, Amsterdam. doi: 10.1016/B978-0-444-59576-8.00027-8.

Singh, D. P. (2004). Use of reduced dose of fungicides and seed treatment with *Trichoderma viride* to control wheat loose smut. *Journal of Mycology and Plant Pathology,* 34:396–397.

Singh, V., Upadhyay, R. S., Sarma, B. K. and Singh, H. B. (2016). *Trichoderma asperellum* spore dose depended modulation of plant growth in vegetable crops. *Microbiological Research,* 193:74–86. doi: 10.1016/j.micres.2016.09.002.

Sood, M., Kapoor, D., Kumar, V., Sheteiwy, M. S., Ramakrishnan, M., Landi, M., Araniti, F. and Sharma, A. (2020). *Trichoderma*: the "secrets" of a multitalented biocontrol agent. *Plants,* 9(6):762. doi: 10.3390/plants9060762.

Stefanski, F. S., Camargo, A. F., Scapini, T., Bonatto, C., Venturin, B., Weirich, S. N., Ulkovski, C., Carezia, C., Ulrich, A., Michelon, W., Soares, H. M., Mathiensen, A., Fongaro, G., Mossi, A. J. and Treichel, H. (2020). Potential use of biological herbicides in a circular economy context: a sustainable approach. *Frontiers in Sustainable Food Systems,* 4:521102. doi: 10.3389/fsufs.2020.521102.

Stewart, A. and Hill, R. (2014). Applications of *Trichoderma* in plant growth promotion. In *Biotechnology and Biology of Trichoderma*, edited by V. K. Gupta, M. Schmoll, A. Herrera-Estrella, R. S. Upadhyay, I. Druzhinina and M. G. Tuohy, 415-428. Elsevier, Amsterdam. doi: 10.1016/B978-0-444-59576-8.00031-X.

Tančić-Živanov, S., Medić-Pap, S., Danojević, D. and Prvulović, D. (2020). Effect of *Trichoderma* spp. on growth promotion and antioxidative activity of pepper seedlings. *Braz arch biol technol,* 63. doi: 10.1590/1678-4324-2020180659.

Tang, X.-X., Liu, S.-Z., Sun, Y.-Y., He, F.-M., Xu, G.-X., Fang, M.-J., Zhen, W. and Qiu, Y.-K. (2021). New cyclopentenoneacrylic acid derivatives from a marine-derived fungus *Trichoderma atroviride* H548. *Natural Product Research,* 35(21):3772–3779. doi: 10.1080/14786419.2020.1737053.

Taylor, J. T., Mukherjee, P. K., Puckhaber, L. S., Dixit, K., Igumenova, T. I., Suh, C., Horwitz, B. A. and Kenerley, C. M. (2020). Deletion of the *Trichoderma virens* NRPS, Tex7, induces accumulation of the anti-cancer compound heptelidic acid. *Biochemical and Biophysical Research Communications,* 529(3):672–677. doi: 10.1016/j.bbrc.2020.06.040.

Tchameni, S. N., Ngonkeu, M. E. L., Begoude, B. A. D., Nana L. W., Fokom, R., Owona, A. D., Mbarga, J. B., Tchana, T., Tondje, P. R., Etoa, F. X. amd Kuaté, J. (2011). Effect of *Trichoderma asperellum* and arbuscular mycorrhizal fungi on cacao growth and resistance against black pod disease. *Crop Protection,* 30(10):1321–1327. doi: 10.1016/j.cropro.2011.05.003.

Tiru, Z., Sarkar, M., Pal, A., Chakraborty, A. P. and Mandal, P. (2021). Three dimensional plant growth promoting activity of *Trichoderma asperellum* in maize (*Zea mays* L.) against *Fusarium moniliforme. Archives of Phytopathology and Plant Protection,* 54(13–14):764–781. doi: 10.1080/03235408.2020.1860420.

Tomah, A. A., Alamer, I. S. A., Li, B. and Zhang, J.-Z. (2020). A new species of *Trichoderma* and gliotoxin role: a new observation in enhancing biocontrol potential of *T. virens* against *Phytophthora capsici* on chili pepper. *Biological Control,* 145:104261. doi: 10.1016/j.biocontrol.2020.104261.

Triolet, M., Guillemin, J.-P., Andre, O. and Steinberg, C. (2020). Fungal-based bioherbicides for weed control: a myth or a reality? *Weed Research,* 60(1):60–77. doi: 10.1111/wre.12389.

Tucci, M., Ruocco, M., De Mais, L., De Palma, M. and Lorito, M. (2011). The beneficial effect of *Trichoderma* spp. on tomato is modulated by the plant genotype. *Molecular Plant Pathology,* 12(4):341–354. doi: 10.1111/j.1364-3703.2010.00674.x.

Ulrich, A., Lerin, L. A., Camargo, A. F., Scapini, T., Diering, N. L., Bonafin, F., Gasparetto, I. G., Confortin, T. C., Sansonovicz, P. F., Fabian, R. L., Reichert Júnior, F. W., Treichel, H., Müller, C. and Mossi, A. J. (2021). Alternative bioherbicide based on *Trichoderma koningiopsis*: enzymatic characterization and its effect on cucumber plants and soil organism. *Biocatalysis and Agricultural Biotechnology,* 36:102127. doi: 10.1016/j.bcab.2021.102127.

Wang, Z., Li, Y., Zhuang, L., Yu, Y., Liu, J., Zhang, L., Gao, Z., Wu, Y., Gao, W., Ding, G. and Wang, Q. (2019). A rhizosphere-derived consortium of *Bacillus subtilis* and *Trichoderma harzianum* suppresses common scab of potato and increases yield. *Computational and Structural Biotechnology Journal,* 17:645–653. doi: 10.1016/j.csbj.2019.05.003.

Yin, M., Fasoyin, O. E., Wang, C., Yue, Q., Zhang, Y., Dun, B., Xu, Y. and Zhang, L. (2020). Herbicidal efficacy of harzianums produced by the biofertilizer fungus, *Trichoderma brevicompactum. AMB Express,* 10(1):118. doi: 10.1186/s13568-020-01055-x.

Yuan, S., Li, M., Fang, Z., Liu, Y., Shi, W., Pan, B., Wu, K., Shi, J., Shen, B. and Shen, Q. (2016). Biological control of tobacco bacterial wilt using *Trichoderma harzianum* amended bioorganic fertilizer and the arbuscular mycorrhizal fungi *Glomus mosseae. Biological Control,* 92:164–171. doi: 10.1016/j.biocontrol.2015.10.013.

Zhu, H., Ma, Y., Guo, Q. and Xu, B. (2020). Biological weed control using *Trichoderma polysporum* strain HZ-31. *Crop Protection,* 134:105161. doi: 10.1016/j.cropro.2020.105161.

Zin, N. A. and Badaluddin, N. A. (2020). Biological functions of *Trichoderma* spp. for agriculture applications. *Annals of Agricultural Sciences,* 65(2):168–178. doi: 10.1016/j.aoas.2020.09.003.

Chapter 3

A Green Solution to Maize Late Wilt Disease

Ofir Degani*, PhD

Plant Sciences Department, MIGAL—Galilee Research Institute, Kiryat Shmona, Israel
Faculty of Sciences, Tel-Hai College, Upper Galilee, Tel-Hai, Israel

Abstract

Late wilt disease of maize, caused by *Magnaporthiopsis maydis*, is considered a major threat to commercial fields in Israel, Egypt and other countries. Today's control methods include chemical and agronomical intervention, but they rely almost solely on resistant maize cultivars. Global disease research is focusing on eco-friendly biological approaches to restrain the pathogen. To this end, we tested nine isolates of *Trichoderma* spp. known for their high mycoparasitic potential as biocontrol agents against *M. maydis* in the lab. High-potential biocontrol agents were selected and evaluated in growth room sprouts, in full-season greenhouse plants and in a commercial field. In addition, we isolated and identified the *Trichoderma asperellum*-secreted metabolite, 6-pentyl-α-pyrone, exhibiting strong *M. maydis* antifungal activity. This pure ingredient was tested against the pathogen in plants over a full growth period. Finally, manipulating the plant's endophytes and soil mycorrhiza (by selected crops in rotation with maize and minimal tillage) has a high potential to reduce late wilt disease. The results of these studies suggest a biological-based protective approach that may have significant value in late wilt integrated prevention.

Keywords: *Cephalosporium maydis,* biological control, crop protection, fungus, *Harpophora maydis*, *Magnaporthiopsis maydis, Trichoderma*

* Corresponding Author's Email: d-ofir@migal.org.il.

In: Trichoderma: Taxonomy, Biodiversity and Applications
Editor: Michael S. Mouton
ISBN: 979-8-88697-946-6

Introduction

Zea mays L. (maize, corn) is one of the world's leading crops for food, feed and fuel and as a raw material for different industrial products (Gálvez Ranilla, 2020). Worldwide annual maize production is expanding at a rate of 1.6%. It was predicted that this rate would not meet the global demand in 2050 (Ray et al., 2013). Among many diseases threatening this cultivar (Pratap and Kumar, 2014; Mueller et al., 2020), late wilt disease (LWD) has been reported so far in ca. ten countries (Figure 1) and is considered a major concern in highly infected countries such as Egypt (El-Naggarr et al., 2015), Israel (Degani and Cernica, 2014), India (Sunitha et al., 2020), Spain and Portugal (Ortiz-Bustos et al., 2015). Economic losses due to LWD were up to 40% in Egypt (Samra et al., 1971), 50-100% in Israel (Drori et al., 2013; Degani et al., 2019b) and 51% in India (Payak and Sharma, 1978). Incidences of the disease can reach up to 100% in Egypt and Israel, and 70% in India. Although the disease has not been reported in the United States, its causal agent, *Magnaporthiopsis maydis*, is regarded as a potentially high-risk phytopathogen (Johal et al., 2004; Bergstrom et al., 2008). LWD harms yield production by erupting at the flowering growth phase, resulting in severe dehydration and plant death.

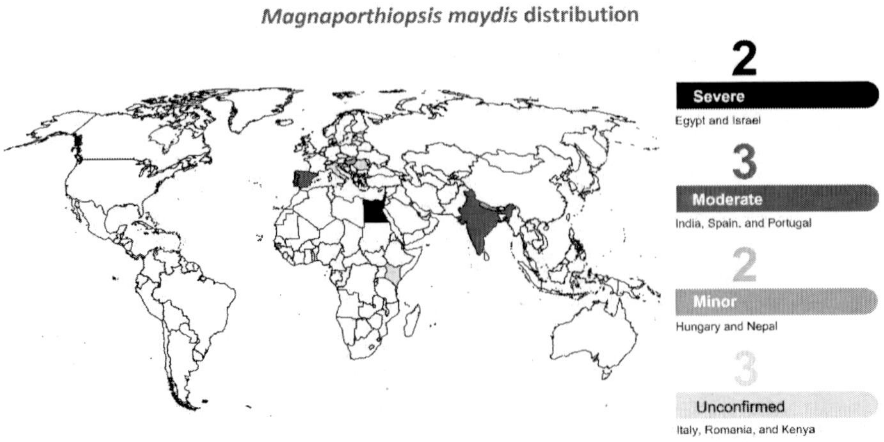

Figure 1. World distribution map of *Magnaporthiopsis maydis*. Disease severity is appraised according to the literature reports and is based on three categories: severe (Egypt and Israel); moderate (India, Spain and Portugal); and minor (Hungary and Nepal). There are also unconfirmed reports of the disease in Italy, Romania and Kenya (adapted from Degani, 2021).

Since the discovery of LWD in Egypt in the early 1960s (Samra et al., 1962), worldwide scientific efforts have led to considerable progress in understanding the disease mode and the pathogen causing it, *M. maydis* (Johal et al., 2004). Moreover, specific research tools for the study of LWD have been developed and applied in the lab, in growth room experiments under controlled conditions and in field trials. A significant part of these efforts was dedicated to creating diverse control methods to restrict the disease's burst and spread and minimize its impact on commercial maize manufacture.

Over the past 60 years, vast efforts dedicated to late wilt disease control have been made (Degani, 2022). The inspected control methods produced different degrees of success and include agrotechnology options (flood fallowing and balanced soil fertility) (Samra et al., 1966b; Singh and Siradhana, 1990), bio-friendly approaches (Elshahawy and El-Sayed, 2018), physical (solar heating) (Fayzalla et al., 1994), allelochemical (Tej et al., 2018), and chemical pesticide (Abd-el-Rahim et al., 1982; Degani and Cernica, 2014; Degani et al., 2014) practices. Recently, the tillage system's impact, the cover crop, and crop rotation have been shown to serve as bioprotective factors against *M. maydis* (Patanita et al., 2020; Degani et al., 2021b; Degani et al., 2022b).

A targeted research effort led to an advancement in our capability to eradicate LWD chemically. A practical, efficient and economic Azoxystrobin-based control protocol (Degani et al., 2014; Degani et al., 2018; Degani et al., 2019b; Degani et al., 2020c) was developed, which can be applied commercially to protect LWD-susceptible maize cultivars. Notwithstanding this recent encouraging achievement, intensive chemical intervention has several short- and long-term drawbacks. In the short term, an intensive chemical application may cause the emergence of resistance to the fungicide. Such situations are becoming more and more common (Massi et al., 2021). In the long-range, phytoparasitic fungi chemical eradication may result in environmental, human and animal risks.

The limitation of chemical fungicides has become critical and is currently a global priority (Ons et al., 2020). Hence, considerable research efforts in the past two decades were dedicated to seeking alternative methods of LWD control. Most of these efforts focused on eco-friendly substitutes to traditional chemical approaches, including the use of *Trichoderma* spp. or other beneficial microorganisms as a biocontrol agent (see, for example, Elshahawy and El-Sayed, 2018; Ghazy and El-Nahrawy, 2020). Late wilt green control studies also aimed at developing soil conservation practices that promote antagonizing mycorrhizal fungi (summarized by Ghazy and El-Nahrawy,

2020). Even though this scientific course has been explored extensively against many harmful plant pathogens (Sood et al., 2020), substantial knowledge gaps exist in regard to LWD. Consequently, the potential of green approaches to control *M. maydis* has only now been revealed.

Currently, the most eco-friendly, cost-effective and efficient method to restrict *M. maydis* is by using highly resistant maize varieties (Rakesh et al., 2016; Gazala et al., 2021). Yet, the discovery of *M. maydis* highly aggressive isolates (Zeller et al., 2002; Ortiz-Bustos et al., 2015; Agag et al., 2021) is a constant problem. These fungal strains may threaten resistant maize cultivars, especially when growing resistant cultivars for extended periods in the same location. Such a scenario may lead to gradual LWD susceptibility weakening (Degani et al., 2014; Degani et al., 2019a). This alarming situation is pushing researchers to continue to seek new methods to control LWD.

Late Wilt Disease

The late wilt causal agent, *M. maydis*, is a seed-borne and soil-borne vascular wilt fungal pathogen that penetrates the host roots and colonizes the xylem tissue (Sabet et al., 1970; Michail et al., 1999). The taxonomic tree of this fungus is phylum: *Ascomycota*, subphylum: *Pezizomycotina*, class: *Sordariomycetes*, subclass: *Sordariomycetidae*, family: *Magnaporthaceae*, genus: *Magnaporthiopsis*, species: *Magnaporthiopsis maydis*. Former scientific names are *Cephalosporium maydis* (Samra, Sabet & Hing, 1963) (Samra et al., 1963) and *Harpophora maydis* (Samra, Sabet & Hing, 1963; Gams, 2000).

M. maydis spreads as sclerotia, spores, or hyphae on the plants' residues (Sabet et al., 1970). The pathogen can persist in the maize stubble and remains; no-till systems may help preserve it (Johal et al., 2004). *M. maydis* can survive in the ground for lengthy periods or by thriving inside diverse host plants, such as lupine (*Lupinus termis* L.) (Sahab et al., 1985), cotton (*Gossypium hirsutum* L.) (Sabet et al., 1966; Degani et al., 2020b), watermelon (*Citrullus lanatus*) and green foxtail (*Setaria viridis*) (Dor and Degani, 2019; Degani et al., 2020b; Degani et al., 2022a).

The disease mode in LWD-sensitive maize cultivars is well-detailed in the scientific literature (Figure 2). *M. maydis* infects maize seedlings during the first three weeks from sowing through their roots or mesocotyl (the seed-coleoptile connecting tissue). It may occasionally cause seed rot or pre-emergence damping-off under high inoculum pressure (Samra et al., 1966a).

At the sprouting phase, *M. maydis* may disrupt root development (Samra et al., 1962) and cause the appearance of necrotic lesions on the roots (Tej et al., 2018). As the plants grow, they are less infected and become LWD-resistant about 50 days after sowing (Sabet et al., 1970). After root penetration, *M. maydis* colonizes xylem tissue (identified 21 days after sowing) and is rapidly transferred to the upper parts of the plant.

The second critical infection phase starts when tassels first emerge (ca. day 55-65, R1 silking, silks visible outside the husks). At this stage, the fungus hyphae and conidia appear throughout the stalk (Sabet et al., 1970), pathogen DNA levels reach their highest point in the stems (Drori et al., 2013) and the first aboveground symptoms are revealed. Later, when *M. maydis* colonizes the entire stalk, a vascular tissue occlusion by hyphae and gum-like secreted materials occurs, resulting in water supply suffocation, rapid dehydration and death (Sabet et al., 1970; Johal et al., 2004). Although the disease appears as patches scattered in the field in many cases (Degani et al., 2019a), LWD may result in total field infection and yield loss in heavily infected areas planted with susceptible maize cultivars (Degani et al., 2014; Degani et al., 2019b). A parallel asymptomatic infection mode, with some delay, occurs in resistant cultivars. This process can result in infected seeds that enhance the pathogen spread (Drori et al., 2013; Degani et al., 2019b).

Maize Late Wilt Biological Control

Since fungicide treatment limitation is exerting increasing pressure in many countries due to environmental and public health concerns, searching for environmentally friendly alternatives to cope with LWD is an ongoing worldwide effort. Many studies were directed towards LWD biological control [20,31,47-49] to address this challenge. These methods include operating and strengthening beneficial microorganism communities in the soil (for example, by compost addition (El-Moghazy et al., 2017)) or direct intervention using antagonistic bacteria and fungi or their secreted metabolites.

Trichoderma spp. Maize Late Wilt Biocontrol

Late wilt disease can be controlled biologically using *Trichoderma* spp. This genus' species can form endophytic mutualistic relationships with various plant species (Harman et al., 2004). Other *Trichoderma* species have been identified to possess biocontrol potential against the plants' fungal pathogens (Harman, 2006). The potential for using *Trichoderma*-based treatment against Israeli *M. maydis* strains has only recently been tested (Degani and Dor, 2021). Examining nine marine (Gal-Hemed et al., 2011) and soil isolates of *Trichoderma* spp. known for their high mycoparasitic potential, revealed that *Trichoderma longibrachiatum* (T7407) and *Trichoderma asperelloides* (T203) isolates exhibited solid antagonistic activity against the Israeli *M. maydis* strain. These eco-friendly agents were tested in a series of experiments in the laboratory (Figure 3) and in a growth room under controlled conditions until their final examination in pots under field conditions throughout an entire growing season (Degani and Dor, 2021). The *T. longibrachiatum* (T7407) green treatment significantly improved growth and yield indices to healthy plant levels, reduced pathogen DNA in the plants' tissues by 98% and prevented disease symptoms (Figure 4).

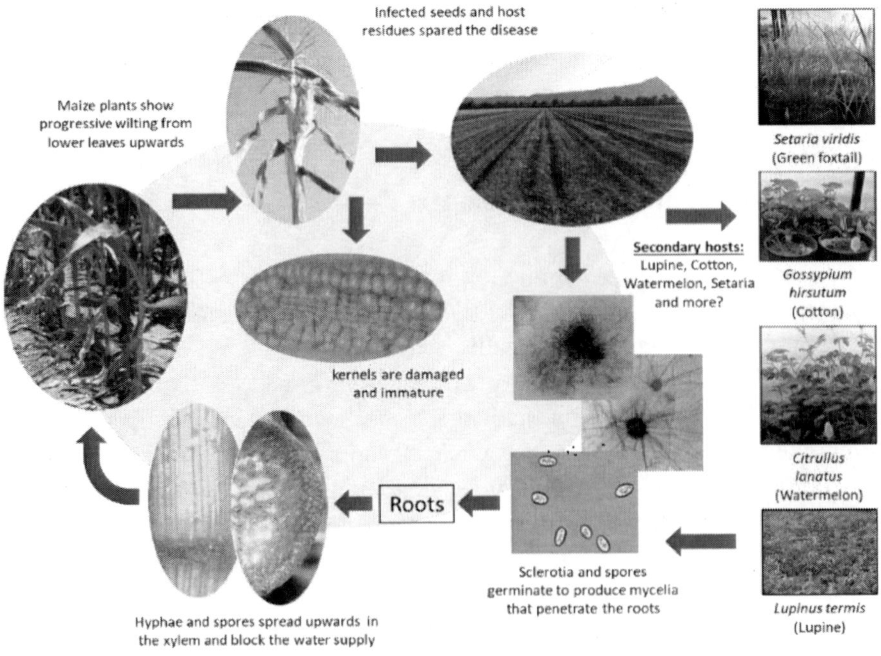

Figure 2. Disease cycle of maize late wilt caused by *Magnaporthiopsis maydis* (adapted from Degani, 2021).

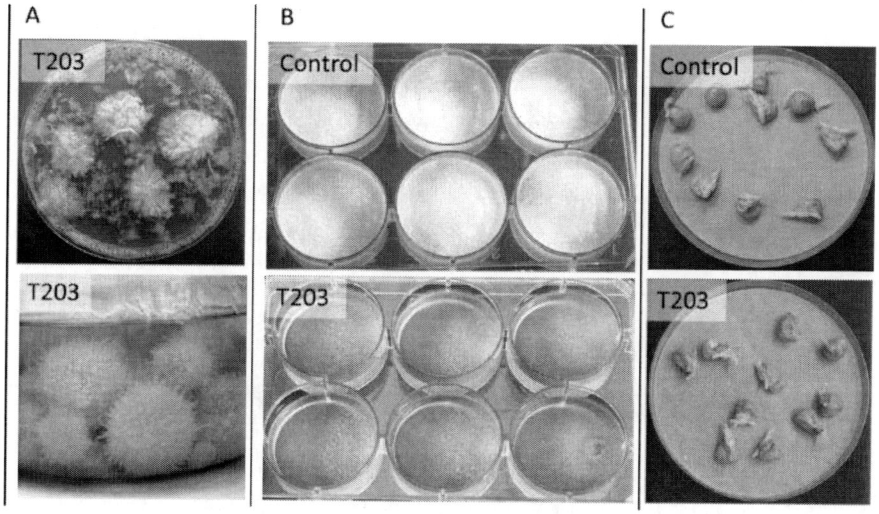

Figure 3. *In vitro* estimation of *Trichoderma asperelloides* (T203)-secreted metabolites-based biological control against *Magnaporthiopsis maydis* (adapted from Degani and Dor, 2021). (**A**) T203-submerged cultures grown by shaking (150 rpm) to isolate secreted metabolites. (**B**) Static shallow media cultures of *M. maydis* on potato dextrose broth (PDB) medium containing T203-secreted metabolites filtrate. Control is PDB medium *M. maydis* cultures maintained under the same conditions. (**C**) Effect of growth media of T203 isolate on corn seed germination. The seeds were germinated in Petri dishes soaked in 4 mL of PDB (control) or PDB + secretion products (growth medium filtrate six days after T203 growth). All images are displayed after 5-6 days of incubation at 28 ± 1°C in the dark.

Another *Trichoderma* species revealed control ability against the maize pathogen is *Trichoderma asperellum* (strain P1), an endophytic symbiont isolated from the seeds of an LWD-sensitive maize cultivar (Degani et al., 2021a; Degani and Dor, 2021). Adding *T. asperellum* directly to seeds with sowing provides significant protection to sprouts (up to 42 days) in a growth room, with more than two-fold growth promotion and reduced pathogen root infection (detected by real-time PCR). The same procedure applied in a commercial field was less beneficial in rescuing the plants' growth and yield. Still, it reduced the cobs' symptoms by 11% and resulted in nine-fold lower levels of the pathogen's DNA in the stem tissue.

In solid and submerged media culture growth assays, the *Trichoderma* species can secrete soluble metabolites that inhibit or kill the maize pathogen (Figure 3B) (Degani and Dor, 2021). Such a metabolite, 6-pentyl-α-pyrone (6-PP, Figure 5), was recently isolated and identified in the *T. asperellum* growth medium (Degani et al., 2021c). The 6-PP metabolite was previously identified as one of the key bioactive compounds of several *Trichoderma* species (Hamrouni et al., 2020). The *T. asperellum*-purified 6-PP compound (30 μg/seed) was used in seed coating and tested against the *T. asperellum* secretory metabolites' crude (diluted to 50%). At the season's end, these treatments improved plant biomass by 90-120% and cob weight by 60%. Moreover, the treatments significantly ($p < 0.05$) reduced the symptoms (up to 20%) and pathogen infection (94-98%) (Degani and Gordani, 2022) (Figure 6).

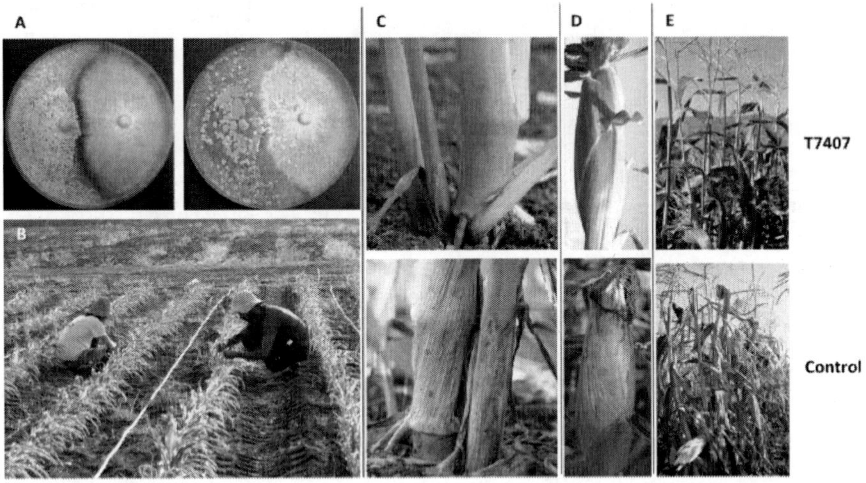

Figure 4. *Trichoderma longibrachiatum* (T7407) biological control against *Magnaporthiopsis maydis* in the lab and in the field (adapted from Degani and Dor, 2021; Degani et al., 2021d). (A) Plate mycoparasitism assay to identify interactions between *M. maydis* and T7407 in a potato dextrose agar (PDA)-rich medium. The two fungi were placed opposite each other, T7407 on the left and *M. maydis* on the right. Photos were taken after three and ten days of growth. (B) Field inoculation of 20-day-old seedlings by an *M. maydis*-infected toothpick. The toothpicks were used for stabbing each plant at the near-surface portion of the stem. (C) The lower stem (first aboveground internode) disease symptoms. (D) The cobs' spathes disease symptoms. (E) The experiment's plots. Representative images of the field plants were taken 82 days after sowing. Controls are unprotected diseased plants.

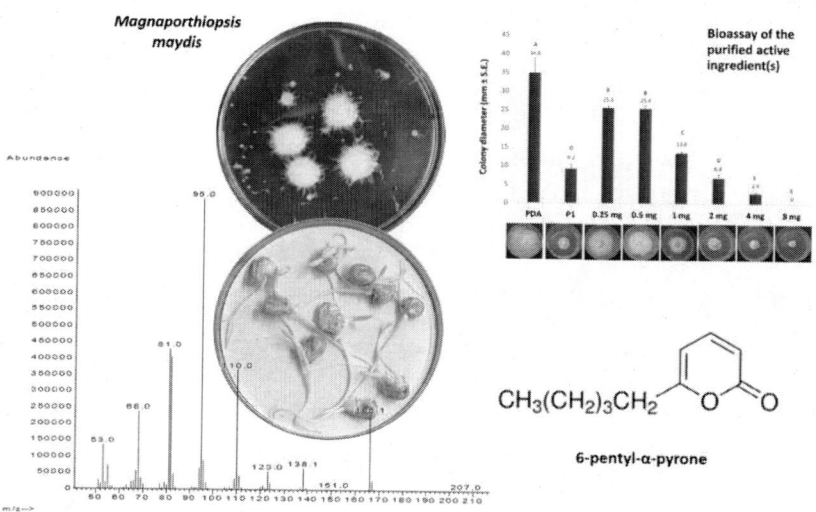

Figure 5. Examination and identification of the *Trichoderma asperellum* (P1) active ingredient using GC-MS analysis. The potent *Magnaporthiopsis maydis* antifungal metabolite 6-pentyl-α-pyrone is secreted by *T. asperellum* (P1), an endophyte isolated in the laboratory from maize seeds of a cultivar susceptible to LWD (Degani et al., 2021a). This purified active compound was tested in a bioassay in solid growth medium cultures and seeds (adapted from Degani et al., 2021c).

Manipulating the Plant Microbiome

The natural microorganisms' communities inhabiting the plant's phyllosphere (the plant's aboveground habitat) or the rhizosphere (the roots' nearby habitat) comprise pathogenic and non-pathogenic members that may protect the plant from pathogens. The soil pathogen, *M. maydis*, interacts with the maize endophytes, which may provide the plant's first defense line. Recently, such endophytes were isolated from six sweet and fodder maize hybrids with different sensitivities to LWD (Degani et al., 2021a). These included ten fungal species and one bacterial species, *Bacillus subtilis* (Figure 7). The fungal species belonged to *Chaetomium*, *Trichoderma*, *Penicillium*, *Rhizopus*, *Alternaria* and *Fusarium* genera. Some are plant pathogens (*Alternaria alternata* and *Fusarium proliferatum*), so they play complex roles in their interactions with the host plant. Enriching seeds with either *Chaetomium subaffine* or *T. asperellum* significantly promoted the infected plants' growth parameters 42 days past sowing. The fungal species *Chaetomium cochliodes*,

T. asperellum, Penicillium citrinum, and the bacteria *B. subtilis* treatments reduced the LWD pathogen DNA in the host plant's roots (Degani et al., 2021a).

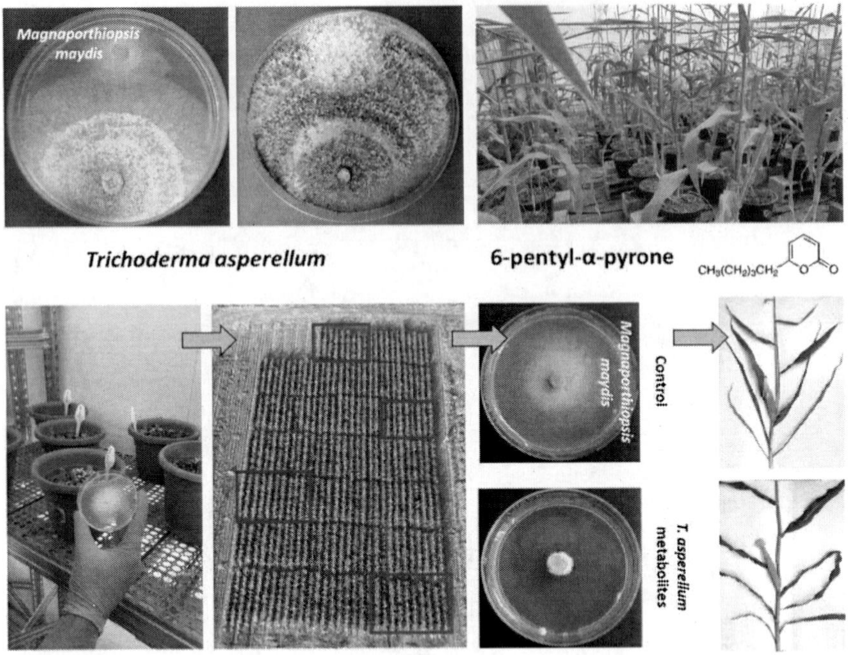

Figure 6. Developing a biological environmentally friendly control strategy against *Magnaporthiopsis maydis* LWD based directly on *Trichoderma asperellum* (P1), its secreted metabolites' crude, or the pure antifungal compound, 6-pentyl-α-pyrone. To this end, growth room trials in seedlings (up to 42 days) and full-season experiments in a net house or the field were conducted (adapted from Degani and Gordani, 2022).

Even potential pathogens can act as beneficial bioprotective agents. In cotton plants, interactions between *M. maydis* and *F. oxysporum* (the cotton wilt agent) led to an interesting result—reduced symptoms of the cotton wilt disease (Sabet et al., 1966). The infection reduction was maximized when *M. maydis* preceded *F. oxysporum* in the soil compared to simultaneous inoculation. Co-infection of the plants with both fungi resulted in immunity compared to infection with *F. oxysporum* alone. In contrast, there was little or no protective outcome when *M. maydis* was added to the soil after *F. oxysporum* (Sabet et al., 1966). Recently, similar antagonistic interactions were identified between *M. maydis* and *Macrophomina phaseolina*, the cotton charcoal rot disease agent (Degani et al., 2020b; Degani et al., 2022a).

Figure 7. Isolation of LWD protective endophytes from maize seed. Maize seeds were externally disinfected, cut lengthwise, placed on a rich substrate and incubated at 28°C in the dark for 2-3 days. At the end of this period, the fungal and bacterial endophytes were isolated and identified. Plate confrontation assays were conducted on all isolates (see example photo of *Bacillus subtilis* against *Magnaporthiopsis maydis* after six days). Potential bioprotective endophytes were applied to enrich the seeds. The treated seeds were planted in highly infected soil, and sprout growth and infection were evaluated on day 42. Background – aerial photographs from 2017 of a maize late wilt-diseased commercial field in northern Israel (adapted from Degani et al., 2021a with modifications).

Beneficial Mycorrhizal Communities

Preserving soil mycorrhizal fungi between growth periods has been essential in crop protection (summarized by Ghazy and El-Nahrawy, 2020). Results from other crop diseases suggest that no-till cropping, crops in a rotation or selected cover crops could support building mycorrhizal communities that function through a sequence of several crops (Brígido et al., 2017). Intensive tillage combined with extended periods where the field is unprocessed results in damaging the integrity of mycorrhizal networks. Preserving the integrity and continuity of the soil mycorrhizal networks may provide the plant with higher resistance to soil diseases [73], including late wilt disease (Patanita et al., 2020). Indeed, arbuscular mycorrhizal fungi (AMF) can improve plants' resistance to biotic and abiotic stresses. This results from activating the plant's local and systemic defense mechanisms (Diagne et al., 2020). Still, a lack of information exists on maize performance under LWD stress in reduced tillage and crop rotation.

In Israel, agricultural practice based on conserving soil microflora integrity (by avoiding tillage) and influencing its nature (by cultivating specific crops in a dual-season growth cycle) was applied in two subsequent studies (Degani et al., 2021b; Degani et al., 2022b). The first study demonstrated that wheat-maize rotation while avoiding tillage can provide some defense against the pathogen *M. maydis*. Still, this preliminary work left many unanswered questions, particularly about the effect of this approach in other maize hybrids and under commercial field conditions. To meet this challenge, the second study applied these agrotechnical practices in parallel as dual-crop experiments—in the greenhouse and in the field—using a hyper-susceptible maize hybrid (Megaton cv.). The clover-maize rotation, especially in the no-till soil, was superior to the other practices tested (the wheat-maize cycle and the commercial Resid MG preparation), and was reflected in growth promotion and LWD durability. The wheat-maize sequence was successful in the greenhouse but ineffective in the field. Surprisingly, the commercial Resid MG soil treatment, which had no evident advantage in the greenhouse, resulted in impressive plant health promotion in the tillage plots in the field. To conclude from these studies and other conducted abroad, choosing the specific crop cycle, cover crop and tillage system could reduce LWD pressure and assist in restricting fungicide use, which has adverse risks to the environment and human health. Since different maize cultivars may react differently to such agrotechnical practices, each cultivar should be tested separately to determine its specific LWD protection suite. Field studies and long-term studies are needed to fully understand the potential of this approach.

Future Challenges and Opportunities

Late wilt of maize is a challenging disease that is imposing a significant economic price in infected areas. The primary method of LWD restriction, the use of resistance germline, is an efficient, economical and environmentally friendly solution that should be maintained and improved. Alongside this primary purpose, the continued development of other solutions to efficiently control the LWD pathogen *Magnaporthiopsis maydis* for commercial maize production is urgently needed in Israel, Egypt, Spain, India and other countries (Galal et al., 1979; Ortiz-Bustos et al., 2015; Degani et al., 2020a; Sunitha et al., 2020). Chemical control of LWD is considered the most effective solution in highly infected areas to protect susceptible maize cultivars. Azoxystrobin-based commercial mixtures can be applied at a timetable adjusted to *M. maydis* pathogenesis and shield-susceptible maize cultivars, even in a highly infected area (Degani et al., 2014; Degani et al., 2018; Degani et al., 2020c). Yet, such a method depends on dripline irrigation applied to couples of rows, which is not suitable in many maize-growing areas and is less economical. In addition, intensive chemical treatment has several short-term and long-term drawbacks. In the short term, an intensive chemical application may result in fungicide resistance. Such cases have gradually become more common (Massi et al., 2021). In the long term, fungicides that restrain plant parasitic fungi may lead to human, animal and environmental risks.

Thus, future efforts devoted to this direction should focus on searching for new hazard-free chemicals that are highly effective against *M. maydis*. Varied effective application techniques of successfully tested pesticides should be developed to meet the growers' needs. There is also a strong necessity to minimize pesticides' dosages and produce and inspect blends of fungicides to prevent the emergence of fungal resistance lines.

One option that could solve many of the problems posed by chemical crop protection is combining chemical and biological approaches (Ons et al., 2020). This solution has been proposed to reduce fungicide doses (and their residues' impact on harvested crops). In addition, combining antifungal chemical and biological treatments reduces the selection pressure on pathogens and, thereby, the chances of resistance development. However, many knowledge gaps need to be addressed to enable this implementation method. A pioneering research work published lately (Gordani et al., 2023) demonstrated the potential of such LWD integrated control interphase and is an encouraging starting point for future field studies that would establish the method.

Today, biopesticides have gained considerable attention in the scientific world because they are a vital alternative to traditional chemical pesticides used to protect field crops. While the development of new eco-friendly options is at the forefront of many novel research studies, there is still a requirement for improved existing protocols, as demonstrated by Elshahawy and El-Sayed

(2018). Their work suggests maximizing the efficacy of *Trichoderma* against *M. maydis* using freshwater microalgae extracts. Such a research direction is opening the door for many similar solutions that, if adequately developed and tested, may produce highly effective and economical solutions to LWD.

Bio-friendly protective microorganisms produce secondary metabolites exhibiting strong antifungal activity. Such a metabolite is 6-pentyl-α-pyrone (6-PP) (Degani et al., 2021c; Degani and Gordani, 2022). This ingredient plays a pivotal role in the biological control of several important phytopathogens and thus may provide a broad defense to the plant. The clean *T. asperellum* 6-PP-secreted product exhibits a highly effective antifungal activity against *M. maydis*. These isolation, identification and examination studies are the first steps in discovering its commercial potential as a new fungicide. Future field studies should confirm this purified component's effectiveness as a seed coating or other preventive treatments to shield highly susceptible maize hybrids against LWD.

Any *in vitro* work to identify new metabolites with *M. maydis* antagonism must be followed by *in vivo* work, which would include *M. maydis*-infected seedlings and mature plants over an entire growth period under field conditions. These further steps, the sprouts and the field trials, would ultimately enable reaching a concluding decision—will the metabolite be practical on a commercial field scale? If so, an identified and validated chemical or mixture of several materials having high efficiency against the LWD causal agent could be proposed as a new pesticide's main ingredient and commercially developed into a future product.

As recently shown, one important source of potential microorganisms exhibiting powerful activity against *M. maydis* is the maize plants themselves (Degani et al., 2021a). The roots or seeds of maize plants (apparently LWD-susceptible cultivars are preferred) are inhabited by many beneficial fungi and bacteria that shield the plant from outside invading pathogens. Identification of these members of the plant microbiome and exploring their potential may open a vast array of new possibilities to control *M. maydis*. Therefore, we are encouraged to widen and deepen our understanding to reveal the true potential of the maize microbiome in the plant survival struggle against the pathogen. A better understanding of these interactions under natural conditions will help us understand, influence and take advantage of the endophyte-based biocontrol potential.

References

Abd-el-Rahim M F, Sabet K A, El-Shafey H A, El-Assiuty E M. Chemical control of the late-wilt disease of maize caused by *Cephalosporium maydis*. *Agric. Res. Rev.* (1982) 60:31-49.

Agag S H, Sabry A M, EL-Samman M G, Mostafa MH. Pathological and molecular characterization of *Magnaporthiopsis maydis* isolates causing late wilt in maize. *Egyptian Journal of Phytopathology* (2021) 49:1-9.

Bergstrom G, Leslie F J, Huber D, Lipps P, Warren H, Esker P, et al. Recovery plan for late wilt of corn caused by *Harpophora maydis* syn. *Cephalosporium maydis*. In: Huber MD (ed). USA: National Plant Disease Recovery System (NPDRS) (2008), p. 24.

Brígido C, van Tuinen D, Brito I, Alho L, Goss M J, Carvalho M. Management of the biological diversity of AM fungi by combination of host plant succession and integrity of extraradical mycelium. *Soil Biology and Biochemistry* (2017) 112:237-247.

Degani O. A review: late wilt of maize—the pathogen, the disease, current status and future perspective. *Journal of Fungi* (2021)7:989.

Degani O. Control strategies to cope with late wilt of maize. *Pathogens* (2022) 11:13.

Degani O, Cernica, G. Diagnosis and control of *Harpophora maydis*, the cause of late wilt in maize. *Advances in Microbiology* (2014) 04:94-105.

Degani O, Dor S. *Trichoderma* biological control to protect sensitive maize hybrids against late wilt disease in the field. *Journal of Fungi* (2021) 7:315.

Degani O, Gordani A. New antifungal compound, 6-pentyl-α-pyrone, against the maize late wilt pathogen, *Magnaporthiopsis maydis*. *Agronomy* (2022) 12:2339.

Degani O, Weinberg T, Graph S. Chemical control of maize late wilt in the field. *Phytoparasitica* (2014) 42:559-570.

Degani O, Danielle R, Dor S. The microflora of maize grains as a biological barrier against the late wilt causal agent, *Magnaporthiopsis maydis*. *Agronomy* (2021a) 11:965.

Degani O, Becher P, Gordani A. Pathogenic interactions between *Macrophomina phaseolina* and *Magnaporthiopsis maydis* in mutually infected cotton sprouts. *Agriculture* (2022a) 12:255.

Degani O, Dor S, Movshovitz D, Rabinovitz O. Methods for studying *Magnaporthiopsis maydis*, the maize late wilt causal agent. *Agronomy* (2019a) 9:181.

Degani O, Regev D, Dor S, Rabinovitz O. Soil bioassay for detecting *Magnaporthiopsis maydis* infestation using a hyper susceptible maize hybrid. *Journal of Fungi* (2020a) 6:107.

Degani O, Dor S, Abraham D, Cohen R. Interactions between *Magnaporthiopsis maydis* and *Macrophomina phaseolina*, the causes of wilt diseases in maize and cotton. *Microorganisms* (2020b) 8:249.

Degani O, Gordani A, Becher P, Dor S. Crop cycle and tillage role in the outbreak of late wilt disease of maize caused by *Magnaporthiopsis maydis*. *Journal of Fungi* (2021b) 7:706.

Degani O, Khatib S, Becher P, Gordani A, Harris R. *Trichoderma asperellum* secreted 6-pentyl-α-pyrone to control *Magnaporthiopsis maydis*, the maize late wilt disease agent. *Biology* (2021c) 10:897.

Degani O, Rabinovitz O, Becher P, Gordani A, Chen A. *Trichoderma longibrachiatum* and *Trichoderma asperellum* confer growth promotion and protection against late wilt disease in the field. *Journal of Fungi* (2021d) 7:444.

Degani O, Gordani A, Becher P, Chen A, Rabinovitz O. Crop rotation and minimal tillage selectively affect maize growth promotion under late wilt disease stress. *Journal of Fungi* (2022b) 8:586.

Degani O, Dor S, Movshowitz D, Fraidman E, Rabinovitz O, Graph S. Effective chemical protection against the maize late wilt causal agent, *Harpophora maydis*, in the field. *PLoS One* (2018) 13:e0208353.

Degani O, Movshowitz D, Dor S, Meerson A, Goldblat Y, Rabinovitz O. Evaluating Azoxystrobin seed coating against maize late wilt disease using a sensitive qPCR-based method. *Plant Dis* (2019b) 103:238-248.

Degani O, Dor S, Chen A, Orlov-Levin, V Stolov-Yosef A, Regev D, Rabinovitz O. Molecular tracking and remote sensing to evaluate new chemical treatments against the maize late wilt disease causal agent, *Magnaporthiopsis maydis*. *Journal of Fungi* (2020c) 6:54.

Diagne N, Ngom M, Djighaly P I, Fall D, Hocher V, Svistoonoff S. Roles of arbuscular mycorrhizal fungi on plant growth and performance: importance in biotic and abiotic stressed regulation. *Diversity* (2020) 12:370.

Dor S, Degani O. Uncovering the host range for maize pathogen *Magnaporthiopsis maydis*. *Plants* (2019) 8.

Drori R, Sharon A, Goldberg D, Rabinovitz O, Levy M, Degani O. Molecular diagnosis for *Harpophora maydis*, the cause of maize late wilt in Israel. *Phytopathologia Mediterranea* (2013) 52:16-29.

El-Moghazy S, Shalaby ME, Mehesen A A, Elbagory M. Fungicidal effect of some promising agents in controlling maize late wilt disease and their potentials in developing yield productivity. *Environment, Biodiversity and Soil Security* (2017) 1:129-143.

El-Naggarr A A A, Sabryr A M, Yassin M A. Impact of late wilt disease caused by *Harpophora maydis* on maize yield. *J. Biol. Chem. Environ. Sc.* (2015) 10:577-595.

Elshahawy IE, El-Sayed AE-KB. Maximizing the efficacy of *Trichoderma* to control *Cephalosporium maydis*, causing maize late wilt disease, using freshwater microalgae extracts. *Egyptian Journal of Biological Pest Control* (2018) 28:48.

Fayzalla E, Sadik E, Elwakil M, Gomah A. Soil solarization for controlling *Cephalosporium maydis*, the cause of late wilt disease of maize in Egypt. *Egypt. J. Phytopathology* (1994) 22:171-178.

Gal-Hemed I, Atanasova L, Komon-Zelazowska M, Druzhinina IS, Viterbo A, Yarden O. Marine isolates of *Trichoderma* spp. as potential halotolerant agents of biological control for arid-zone agriculture. *Applied and Environmental Microbiology* (2011) 77:5100-5109.

Galal AA, El-Rouby MM, Gad AM. Genetic analysis of resistance to late wilt (*Cephalosporium maydis*) in variety crosses of maize. *Zeitschrift fur Planzenzuchtung* (1979) 83:176-183.

Gálvez Ranilla L. The application of metabolomics for the study of cereal corn (*Zea mays* L.). *Metabolites* (2020) 10:300.

Gams W. Phialophora and some similar morphologically little-differentiated anamorphs of divergent ascomycetes. *Studies in Mycology* (2000) 45:187-200.

Gazala P, Gangappa E, Ramesh S, Swamy D. Comparative breeding potential of two crosses for response to late wilt disease (LWD) in maize (*Zea mays* L.). *Genetic Resources and Crop Evolution* (2021) 1-7.

Ghazy N, El-Nahrawy S. Siderophore production by *Bacillus subtilis* MF497446 and *Pseudomonas koreensis* MG209738 and their efficacy in controlling *Cephalosporium maydis* in maize plant. *Archives of Microbiology* (2020) 203:1195-1209.

Gordani A, Hijazi B, Dimant E, Degani O. Integrated Biological and Chemical Control against the Maize Late Wilt Agent *Magnaporthiopsis maydis*. *Soil Systems* (2023); 7:1.

Hamrouni R, Molinet J, Dupuy N, Taieb N, Carboue Q, Masmoudi A, Roussos S. The Effect of aeration for 6-pentyl-alpha-pyrone, conidia and lytic enzymes production by *Trichoderma asperellum* strains grown in solid-state fermentation. *Waste and Biomass Valorization* (2020) 11:5711-5720.

Harman G E. Overview of mechanisms and uses of *Trichoderma* spp. *Phytopathology* (2006) 96:190-194.

Harman G E, Howell C R, Viterbo A, Chet I, Lorito M. *Trichoderma* species-opportunistic, avirulent plant symbionts. *Nat Rev Microbiol* (2004) 2:43-56.

Johal L, Huber D M, Martyn R. Late wilt of corn (maize) pathway analysis: intentional introduction of *Cephalosporium maydis*. In: Pathways Analysis for the Introduction to the U.S. of Plant Pathogens of Economic Importance (2004).

Massi F, Torriani SF, Borghi L, Toffolatti, SL. Fungicide resistance evolution and detection in plant pathogens: *Plasmopara viticola* as a case study. *Microorganisms* (2021) 9:119.

Michail, S H, Abou-Elseoud MS, Nour Eldin MS. Seed health testing of corn for *Cephalosporium maydis*. *Acta Phytopathologica et Entomologica Hungarica* (1999) 34:35-42.

Mueller, Daren S., Kiersten A. Wise, Adam J. Sisson, Tom W. Allen, Gary C. Bergstrom, D. Bruce Bosley, Carl A. Bradley, Corn yield loss estimates due to diseases in the United States and Ontario, Canada, from 2016 to 2019. *Plant Health Progress* (2020) 21:238-247.

Ons L, Bylemans D, Thevissen K, Cammue B P A. Combining biocontrol agents with chemical fungicides for integrated plant fungal disease control. *Microorganisms* (2020) 8:1930.

Ortiz-Bustos C M, Testi L, García-Carneros AB, Molinero-Ruiz L. Geographic distribution and aggressiveness of *Harpophora maydis* in the Iberian peninsula, and thermal detection of maize late wilt. *European Journal of Plant Pathology* (2015) 144:383-397.

Patanita M, Campos M D, Félix M d R, Carvalho M, Brito I. Effect of tillage system and cover crop on maize mycorrhization and presence of *Magnaporthiopsis maydis*. *Biology* (2020) 9:46.

Payak M, Sharma R. Research on diseases of maize. In: *PL-480 Project Final Technical Report (April 1969-March 1975)*. New Delhi, India: Indian Council of Agricultural Research (1978), p. 228.

Pratap A, Kumar J. *Alien Gene Transfer in Crop Plants, Volume 2: Achievements and Impacts*. New York, NY, USA: Springer (2014).

Rakesh, B., E. Gangappa, Sonali Gandhi, R. P. Veeresh Gowda, S. Dharanendra Swamy, S. Ramesh, and H. B. Hemareddy. Modified method of screening maize inbred lines to late wilt disease caused by *Harpophora maydis*. *Mysore Journal of Agricultural Sciences* (2016) 50:684-690.

Ray D K, Mueller N D, West P C, Foley J A. Yield trends are insufficient to double global crop production by 2050. *PloS one* (2013) 8:e66428.

Sabet K, Samra A, Mansour I. Interaction between *Fusarium oxysporum f. vasinfectum* and *Cephalosporium maydis* on cotton and maize. *Annals of Applied Biology* (1966) 58:93-101.

Sabet K A, Zaher A M, Samra A S, Mansour IM. Pathogenic behaviour of *Cephalosporium maydis* and *C. acremonium*. *Ann Appl Biol* (1970) 66:257-263.

Sahab A F, Osman A R, Soleman N K, Mikhail M S. Studies on root-rot of lupin in Egypt and its control. *Egypt. J. Phytopathol.* (1985) 17:23-35.

Samra A, Sabet K, Abd-el-Rahim M. Seed transmission of stalk-rot fungi and effect of seed treatment. *Investigations on Stalk-Rot Diseases of Maize in UAR* (1966a), pp. 94-116.

Samra A, Sabet K, Kamel M, Abd El-Rahim M. Further studies on the effect of field conditions and cultural practices on infection with stalk-rot complex of maize. The Arab Republic of Egypt, Min. of Agriculture. *Plant Protection Dept., Bull.* (1971).

Samra A S, Sabet K A, Hingorani M K. A new wilt disease of maize in Egypt. *Plant Dis. Rep.* (1962) 46:481-483.

Samra A S, Sabet K A, Hingorani M K. Late wilt disease of maize caused by *Cephalosporium maydis*. *Phytopathology* (1963) 53:402-406.

Samra A S, Sabet K A, Abdel-Rahim M F. *Effect of soil conditions and cultural practices on infection with stalk rots*. Cairo, Egypt: U.A.R. Ministry of Agric. Government Printing Offices (1966b).

Singh S D, Siradhana BS. Effect of macro and micronutrients on the development of late wilt of maize induced by *Cephalosporium maydis*. *Summa Phytopath* (1990) 16:140-145.

Sood, Monika, Dhriti Kapoor, Vipul Kumar, Mohamed S. Sheteiwy, Muthusamy Ramakrishnan, Marco Landi, Fabrizio Araniti, and Anket Sharma. *Trichoderma*: the "secrets" of a multitalented biocontrol agent. *Plants* (2020) 9:762.

Sunitha N, Gangappa E, Gowda R V, Ramesh S, Biradar S, Swamy D, Hemareddy H. Assessment of impact of late wilt caused by *Harpophora maydis* (Samra, Sabet and Hing) on grain yield and its attributing traits in maize (*Zea mays* L.). *Mysore Journal of Agricultural Sciences* (2020) 54:30-36.

Tej R, Rodríguez-Mallol C, Rodríguez-Arcos R, Karray-Bouraoui N, Molinero-Ruiz L. Inhibitory effect of *Lycium europaeum* extracts on phytopathogenic soil-borne fungi and the reduction of late wilt in maize. *European Journal of Plant Pathology* (2018) 152:249-265.

Zeller K A, Ismael A M, El-Assiuty E M, Fahmy Z M, Bekheet F M, Leslie J F. Relative competitiveness and virulence of four clonal lineages of *Cephalosporium maydis* from Egypt Toward greenhouse-grown maize. *Plant Dis.* (2002) 86:373-378.

Chapter 4

Trichoderma: A Potential Bio-Control Agent for Sustainable Agriculture and the Environment

M. Mousumi Das*
and A. Sabu
Department of Biotechnology and Microbiology, Dr. Janaki Ammal Campus, Kannur University, Kerala, India

Abstract

Agriculture and its allied sectors have been facing stiff challenges in the new millennium. The emergence of plant-parasitic microorganisms in a crop plantation causes many serious diseases. Besides, the excessive use of chemical pesticides to control phytopathogens has led to the destruction of soil structure, microbial diversity, soil fertility, and accumulation of toxic substances in soil. The ideal choice to overcome these issues is the application of eco-friendly biocontrol agents in agricultural products. Among the wide variety of myco-biocontrol agents, *Trichoderma* has received remarkable importance in agricultural niche markets because of its enormous potential in bio-control processes compared to other entomopathogenic fungi. They have been gaining momentum as a potential fungal antagonist in the control of crop diseases in recent decades because of their eco-friendly nature, minimizing the use of synthetic agrochemicals, can combat bacterial and fungal infection, and improving crop productivity through multiple roles. The bio-control treatment with certain strains of *Trichoderma* spp. proved to be successful in controlling various soil-borne and foliar pathogens in

---

* Corresponding Author's Email: mousumidas35@gmail.com.

In: Trichoderma: Taxonomy, Biodiversity and Applications
Editor: Michael S. Mouton
ISBN: 979-8-88697-946-6

many countries. The review throws light on the potentiality of *Trichoderma* sp. as a biocontrol agent and its mechanism in plant disease management.

Keywords: *Trichoderma*, biocontrol agent, antagonism, phytopathogen, bioremediation

Introduction

Crop diseases are to be controlled to maintain the quality and quantity of food, feed, and fiber produced by farmers around the world. Many approaches have been used for the identification and control of pests and plant diseases. Conventional agricultural practices and crop protection often rely heavily on synthetic fertilizers, herbicides, and pesticides to control crop diseases and pests, but the reckless and overuse of chemicals in the farm field has given rise to the drastic reduction of beneficial soil microbes, soil fertility, and reduction in soil organic matter. With the increase in demand for organic foods, effective crop disease control by microbial strains is fast becoming inevitable as the conventional farming system of synthetic chemicals is degrading our environment. In recent years, socioeconomic and safety issues related to the environmental impact of agrochemicals has led to the consideration of bio-inputs in sufficient quantities as a natural approach to regaining the lost vitality of soil and improving plant health.

Organic farming is an emerging scientific field in India that is of paramount use for producing high-yielding breeds with better nutritional value. Such a farmer welfare-centered approach to sustainable agriculture can empower rural farmers with higher income and employment and make balanced development a reality. Moreover, biological control methods can do something effective and control the imbalanced growth of plant pests and phytopathogens and offer great scope for sustainable agriculture.

Biocontrol Agents and Its Importance in Agriculture

Bio-control agents (BCAs) include predators, parasites, and pathogens (fungi, bacteria, oomycetes, viruses, protozoa, and nematodes) that control various plant pathogens in crops, although each organism is highly specific to its host (Usta, 2013). Microbial BCAs have different mechanisms of action to control pests, pathogens, and diseases. Disease suppression can be achieved in several ways, such as parasitism, antibiosis, microbial competition for nutrients and space, biocontrol activated stimulation of plant defence mechanisms, hydrogen cyanide, lytic enzymes, induced systematic resistance (ISR), production of growth hormones, and rhizosphere colonization capability (Sharma et al., 2013). During the past years, the utilization of microbial BCAs as sustainable solutions for the management of plant pathogens in crops has been increased to protect the environment from toxic chemicals and offers great potential for field application.

This approach is being adapted around the world and is highly compatible with sustainable agriculture and effective against a wide range of plant pathogens (Das and Abdulhameed, 2020).

The use of BCAs is a natural method of controlling pests including insects, microbial pathogens, mites, weeds, and crop diseases. Many fungi, bacteria, viruses, etc., play a crucial role in sustainable agricultural development. Nowadays, research and application of bacterial and fungal BCAs as a pest control strategy has gained increasing attention. Compared with other BCAs such as bacteria and viruses, there are some distinct advantages in utilizing fungi. They can infect their target host by direct adhesion to the surface and penetration through the cuticle. The attractiveness of fungal BCAs, in particular, is due to their high reproductive capacity, host specificity, short generation time, adaptability to different environmental conditions, persistence in the field, ability to thrive on inexpensive agro-industrial residues, the higher spectrum of disease management, diverse bio-control mechanisms, rapid production of biomass, dispersal efficiency and of their easy maintenance and production characteristics, which can ensure their survival longer without a host (Jyoti and Singh, 2016; Sandhu et al., 2012; Das and Abdulhameed, 2020).

Role of Fungi as Biocontrol Agent

Fungi possess several characteristics that make them potentially ideal BCAs. Antagonistic fungi are both a feasible system for the management of plant diseases in the crop field with a growing commercial market and a significant model for studies of target host-microbial pathogen interaction (Schrank and Vainstein, 2010). According to Das and Abdulhameed (2020), the main reasons for their broad use are their ability to grow on inexpensive agro-processing residues, diverse bio-control mechanisms, increased target specificity, huge metabolic diversity, relative environmental safety, and rapid production of biomass.

The entomopathogenic fungus can act as a parasite of insects and kills or infects them. Since they are considered natural and eco-friendly agents of mortality, there is worldwide interest in the use and manipulation of fungal BCAs for the control of various plant pathogens in agriculture. Using entomopathogens as bio-pesticides in pest management is called microbial control, which can be a critical part of integrated pest management (IPM) against several pests. It typically causes infection when fungal spores encounter the insect host. Under ideal conditions of optimum temperatures and high relative humidity, fungal spores germinate and breach the host exoskeleton or cuticle through enzymatic degradation and mechanical pressure to enter the insect body or haemolymph. Once inside the body, the fungi multiply, and invade the insect tissues, producing toxins and developing the nutrients in the haemocoel to prevent the immune responses of insects (Usta, 2013; Yashaswini and Sudarsanam, 2017; Das and Abdulhameed, 2020).

Fungal antagonists are used singly or in combination with other biological pest control programmes. Many filamentous fungi have been used as BCAs against a diverse array of pests, insects, weeds, and parasitic nematodes. These include bio-insecticides (*Verticillium lecanii*, *Beauveria bassiana*, *Metarhizium anisopliae*, *Gliocladium* sp., *Nomuraea rileyi*), bio-fungicides (*Trichoderma harzianum*, *Trichoderma asperellum*, *Trichoderma viride*,), bio-herbicides (*Phytophthora*), and bio-nematicides (*Purpureocillium lilacinum*). Several filamentous fungi can be applied in the field as conidia or sporulating mycelium for effective biological control of plant pathogens (Das and Abdulhameed, 2020).

Among the various micro-organisms being used as BCAs, the genus *Trichoderma* (*Ascomycota*, *Hypocreales*) is one of the best-known bio-agent because of its antagonistic activities against phytopathogens. The genus *Trichoderma* produces and secrete a wide range of extracellular hydrolytic enzymes which play a pivotal role in biocontrol activity like degradation of cell wall components of fungal or plant cells, resistance to biotic or abiotic stress agents, growth of hyphae, etc. (Shah and Afiya, 2019). The entomopathogenic fungi belonging to the genus *Trichoderma* has acquired immense importance since several decades, because of their antagonistic behavior against several phytopathogens and growth promotion in crop plants (Hhmau et al., 2015).

Trichoderma: A Versatile Weapon for Plant Pathogens

The genus *Trichoderma* (Division- Ascomycota, Subdivision-Pezizomycotina, Class- Sordariomycetes, Order- Hypocreales, Family-Hypocreaceae) is one of the most explored and effective BCA with a high success rate in the control of major plant diseases. It is a free-living, soil inhabitant belonging to the group Deuteromycetes of fungi. *Trichoderma* has been known since the 1920s for its ability to act as BCAs against phytopathogenic micro-organisms (Samuels, 1996). It has more than 100 species, out of which *T. harzianum, T. viride, T. asperellum, T. brevicompactum, T. longibrachiatum, T. koningii*, etc., are proved effective against soil-borne plant pathogens like *F. oxysporum, P. capsici, Pythium spp., R. solani*, and *Sclerotium rolfsii* (Puyam, 2016).

Trichoderma spp. is the most effective and highly successful bio-agent and has been increasingly used in agricultural markets and the greenhouse industry, where a significant proportion of commercially available bio-fertilizers and bio-pesticides are based on *Trichoderma harzianum. Trichoderma harzianum* Rifai (Ascomycota, Sordariomycetes, Hypocreales, Hypocreaceae) is a safe and effective BCA for several plant fungal diseases (Chaverri et al., 2015). These free-living entomopathogenic fungi are ubiquitous in the soil environment and are being successfully used and commercialized to combat a broad spectrum of parasitic nematodes and fungal pathogens including *Rhizoctonia solani*, *Pythium* spp., *Phytophthora* spp., and *Fusarium* spp. The bio-control efficiency of *Trichoderma* against several phytopathogens was demonstrated for several spice and vegetable crops such as *Trichoderma harzianum* against *Phytophthora* foot rot in black pepper and rhizome rot in ginger (Kumar et al., 2012; Singh et al., 2018); *Trichoderma* against *Phytophthora nicotianae* infections on tomato (La Spada et al., 2020); *Trichoderma* against *Phytophthora infestans* on potato (Yao et al., 2016). Besides, many other species of *Trichoderma* such as *T. asperellum, T. brevicompactum*, *T. atroviride*, and *T. virens* are widely used as BCAs against various plant pathogens (Das and Abdulhameed, 2020; Das et al., 2019). All of these are capable of directly killing pathogens and other parasitic fungi of plants, with a wider host range in diverse ecologies. Most commercial *Trichoderma* based formulations use conidiospores in dry or liquid formulation and are delivered through either seed treatment, soil or furrow drench, foliar spray, bio-priming, seedling root dip, or through drip irrigation (Kumar et al., 2014). The renewed interest in BCAs among agriculture biologists is due to their ecological protection against pests, pathogens, and plant diseases, long-lasting effects, and safety features.

Mechanism of Action of *Trichoderma* sp. in Biological Control of Plant Diseases

Various mechanisms have been suggested for the antagonistic activity of *Trichoderma* spp. against plant pathogenic fungi and pests. These mechanisms mainly rely on competition, colonization, antibiosis, direct mycoparasitism, induced resistance, and production of hydrolytic enzymes (cellulases, proteases, β-1, 3 glucanases, and chitinases) and secondary metabolites (Howell, 2003; Nusaibah and Musa, 2019) (Figure 1).

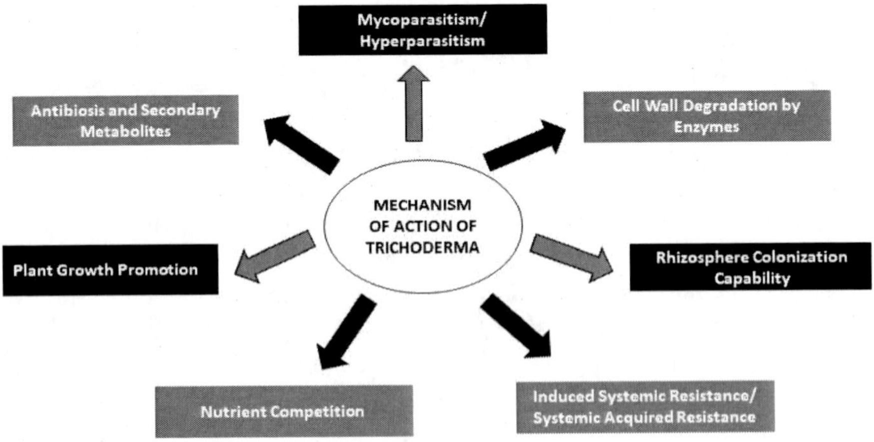

Figure 1. Mechanism of action of *Trichoderma* sp.

Recent reports stated that competition for nutrients is the most important biocontrol mechanism used by *T. harzianum* T-35 to control *F. oxysporum f.sp. vasinfectum* and *F. oxysporum. f.sp. melonis* in cotton and melon under natural soil conditions (Sivan and Chet, 1989). Sariah et al., (2005) demonstrated that *T. harzianum* has a strong competitive ability to displace the causative agent of basal stem rot of oil palm (*Ganoderma boninense*) to reduce the pathogen colonization in the roots.

Trichoderma isolates appear to be an inevitable source of antibiotics. Under nutrient liming conditions, they secrete low molecular weight (<1 kDa) secondary metabolites with fungicidal and bactericidal capabilities, which include peptaibols, gliotoxin, gliovirin, polyketides, terpenoids, 6-pentyl pyrone, siderophores, viridian, harzianolide, and isocyane metabolites (Zeilinger et al., 2016); harzianic acid (Vinale et al., 2013); Epipoly-thiodioxopiperazines, butenolides, pyridones, azaphilones, koninginins, steroids, anthraquinones, lactones, trichothecenes, and other antifungal compounds (Khan et al., 2020b); trichoviridin, trichodermin, sesquiterpene heptalic acid, and dermadin (Nakkeeran et al., 2002). Shakeri and Foster (2007) reported that peptaibiol antibiotics produced by *T. harzianum* have a key role in entomopathogenicity against *Tenebrio molitor* larvae. The bioactive metabolites, harzianolide (P19) and harzianic acid has strong antagonistic activity against the growth of pathogenic fungi such as *Gaeumannomyces graminis var. tritici*, *Pythium irregulare*, *Sclerotinia sclerotiorum*, and *Rhizoctonia solani* (Almassi et al., 1991; Vinale et al., 2009).

Mycoparasitism or hyperparasitism of *Trichoderma* is an important mechanism involved in their antagonistic activities against plant pathogenic fungi. Kumar et al., (1998) studied that *T. harzianum* inhibits the growth of *F. solani* by forming an appressorium on pathogenic hyphae and coiling lightly around them within 95 h of contact. They concluded that within 6 days, the fungus was completely inhibited, while *T. harzianum* was multiplied by conidiogenesis. Naher (2014) investigated that *Trichoderma harzianum* is a potential BCA against a devastating fungus *Ganoderma boninense*, a causative agent of basal stem rot disease in oil palm. The antagonistic activity in *Trichoderma* showed the mycoparasitic process to kill the fungal pathogen. Similarly, the hyperparasitism process was found in *Trichoderma* sp. on *Rhizoctonia solani* (Brotman et al., 2010). Apart from the antagonistic activity of *Trichoderma* spp. against plant pathogens and pests, several reports have shown the induction of systematic and local resistances against a wide range of phytopathogens (Shoresh et al., 2010).

Trichoderma species were efficient BCAs and producers of powerful hydrolytic enzymes. Hydrolytic enzymes such as glucanase, chitinase, and protease can suppress a wide range of phytopathogens (De Marco et al., 2000). El-Katatny et al., (2000) reported that *T. harzianum* culture filtrates exhibited chitinase and β- 1, 3- glucanase activity and could hydrolyze the mycelia of the phytopathogenic fungus *S. rolfsii*. Similarly, Parmar et al., (2015) investigated *Trichoderma* spp. inhibited the growth of the *S. rolfsii* pathogen in groundnut by producing chitinase and protease enzymes.

Myco-parasitic *Trichoderma* spp. can recognize the signals from host hyphae, triggering coiling, develop haustoria and penetrate and degrade the host cell wall with cell degrading enzymes and utilize the content of the host hyphae as a nutrient source (Chet, 1987).

Members of the fungal genus *Trichoderma* produce and secrete a wide range of cell wall-degrading enzymes including chitinase, β-1, 3-glucanase, protease, pectinase, mannanase, cellulase, xylanase, lipase, amylase, and arabinase (Strakowska et al., 2014). Also, secondary metabolites produced by *Trichoderma* confer the biocontrol potential of the species. The most important fungal metabolites produced by *Trichoderma* include gliotoxin, viridin, viridiofungins, nitrogen heterocyclic compounds, trichodenones and cyclopentenone derivatives, alpha-pyrones, terpenes, anthraquinones, daucane, koinginins, trichodermamides, polyketides, isocyano metabolites, piperacines, peptaibols, heptelidic acid, and derivatives, azaphilones, harzialactones and derivatives, butenolides, trichothecenes, statins, setin-like metabolites (trichosetin), bisorbicillinoids, diketopiperazines, ergosterol derivatives, and cyclonerodiol derivatives. All these metabolites produce synergistic effects in combination with cell-wall degrading enzymes with strong antagonistic activity against many phytopathogenic fungi, bacteria, and yeasts (Monte, 2001; Vizcaino et al., 2005; Vinale et al., 2006; Patil and Chakranarayan, 2016).

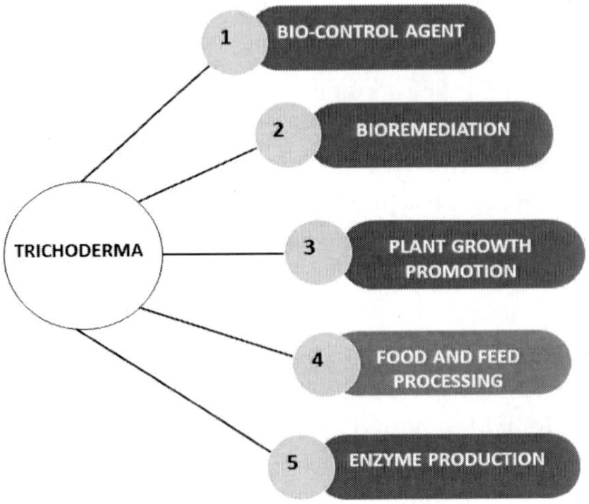

Figure 2. Applications of *Trichoderma* sp. in different field.

The bio-control treatment with certain strains of *Trichoderma* spp. proved to be successful in controlling various soil-borne and foliar pathogens in many countries. In addition to their important role as potential BCA, there are some other applications of *Trichoderma* in various fields, as shown in Figure 2. Field application of *Trichoderma* spp. is mainly achieved through fungal conidiospores or sporulating mycelium, which must be virulent and viable for a stock long time (Das and Abdulhameed, 2020).

Trichoderma: A Beneficial Biocontrol Agent for Eco-Sustainable Agriculture

Trichoderma spp. is having a key role in the management of crop diseases through biological means, because of its ability to combat fungal infection and improve crop productivity through multiple roles (Das and Abdulhameed, 2020). It is one of the most researched genera, able to antagonize and control a wide range of phytopathogens and pests and contributing to as high as 50% of commercially available BCAs fungi. *Trichoderma* species are known producers of secondary metabolites with agricultural significance as they often exhibit antifungal and antibacterial properties. The phenomenon of bio-control employed by *Trichoderma* is driven by either direct antagonism, such as myco-parasitism/hyperparasitism, antibiosis, competition and rhizosphere competence, and sclerotia colonization and parasitization, or indirectly by promoting plant growth and vigor and enhancing tolerance to abiotic stresses, improve nutrient absorption and bioremediation of contaminated soil, as well as providing plants several secondary metabolites, enzymes, and pathogenesis-related (PR) proteins plant defensive mechanisms (Sharma, 2018; Kumar, 2013). These fungi can also enhance the vegetative growth of plants and the nutrient content of soil through decomposition and biodegradation (Shah and Afiya, 2019). The fungal conidiospores (active substance) can be applied to crop plants as foliar sprays and pre-planting and post-pruning treatments, during irrigation and transplanting. Currently, many *Trichoderma-based* bio-formulations are available in the agricultural market globally which are dominated by *T. harzianum* and *T. viride* (Sharma, 2018). It is a safe and eco-friendly method to reduce the detrimental effects of agrochemicals. Various research articles reported on the role of *Trichoderma* spp. as a BCA against various pests and phytopathogens such as *T. harzianum* and *T. viride* against *Alternaria alternata, Drechslera tetramera* in capsicum

(Kapoor, 2011); *T. harzianum* against soil-borne phytopathogen, *R. solani* in cowpea (Pan and Das, 2011); *T. viride* and *T. harzianum* were effective BCAs of *Fusarium udum* in pigeon pea (Hukma and Pandey, 2011); *T. harzianum* was an antagonist to root rot disease of pepper plants (Ahmed et al., 2003); *T. harzianum* controls the *Pythium* sp. causing damping-off in cucumber (Paulitz et al., 1990); *T. asperellum* inhibit and control the growth of *F. oxysporum* f. sp. *lycopersici in tomato (*El Komy et al., 2015); *Trichoderma* spp. significantly inhibited the bacterial growth of *Ralstonia solanacearum* and caused significant juveniles mortality and inhibition in egg hatching of *M. incognita* (Khan et al., 2020); *T. harzianum* against *M. javanica* infesting tomato (Naserinasab et al., 2011); *Trichoderma spp.* reduced 50% of *M. incognita* population (Mascarin et al., 2012); *T. viride*, *T. virens*, and *T. harzianum* against fruit rot causing *R. solani* in tomatoes (Amin and Razdan, 2010); *T. gamssii* against the head blight of wheat and other small cereal grains (Matarese et al., 2012); *T. atroviride* against *S. sclerotiorum (*Li et al., 2005*). Trichoderma* species not only suppress the growth of pathogenic microbes in plants but also improve plant defense responses, and stimulate plant growth (Bastakoti et al., 2017).

Agricultural Significance of *Trichoderma harzianum*

The future of crop production and ecological safety is in jeopardy because of the problems in agriculture. The emergence of plant-parasitic microorganisms in a crop plantation causes many serious diseases. Besides, the excessive use of chemical pesticides to control phytopathogens has led to the destruction of soil structure, microbial diversity, soil fertility, and accumulation of toxic substances in soil. Another problem is the improper management of agro-processing residues, which pollute the surrounding environment when it has been burned or disposed of into water bodies (Shah and Afiya, 2019; Zin and Badaluddin, 2020). The ideal choice to overcome these issues is the application of eco-friendly BCA such as *Trichoderma* spp. in agricultural products. *Trichoderma* spp. could directly inhibit the growth of soil-borne pathogens and regulate the growth rate of plants. Recent studies have demonstrated that several crop diseases such as root rot disease, damping off, sheath blight, web blight, brown spot, wilt, fruit rot, and other plant diseases can be controlled by *Trichoderma* spp. (Zin and Badaluddin, 2020). Among the fungal BCAs, *T. harzianum* is the most explored soil fungi for the control of foliar and soil-borne microbial plant pathogens. Over the past several decades, numerous strains of *T. harzianum* have been isolated from the rhizosphere soil and plant roots by several researchers, and their antagonistic activity against phytopathogens were reported. *T. harzianum* controlled two soil-borne pathogens i.e., *S. rolfsii* and *R. solani* effectively under both greenhouse and field conditions (Elad et al., 1980). Tjamos et al., (1992) reported that *T. harzianum* T-35 inhibits the growth of *F. oxysporum* by competing for both rhizosphere colonization and nutrients, the antagonism becoming more effective as the nutrient concentration decreases. Uddin et al., (2018) indicated that *T. harzinum* had strong antagonistic activity against *Pythium ultimum* and *P. capsici*. Bastakoti et al., (2017) evaluated the *in vitro* efficacy of newly isolated *Trichoderma* spp. in managing various soil-borne phytopathogens such as *Sclerotium rolfsii, R. solani, S. sclerotiorum,* and *Fusarium solani*. Dual culture results over radial growth of *Sclerotium rolfsii* was found to be maximum in comparison to other pathogens. The analyses of growth inhibition assay of all the *Trichoderma* isolates were suggestive of their potential use in the agricultural field to protect plants affected by various fungal pathogens, especially *Sclerotium rolfsii.*

The bio-control potential serves as the base for effective field applications of *T. harzianum* as bio-fungicide or bio-pesticide against soil-borne, foliar, and vascular pathogens and insect pests, as an alternative to agrochemicals to control a wide spectrum of phytopathogens, as well as to increase resistance to biotic or abiotic stress conditions (Bharti et al., 2012).

The global biopesticides market is growing with a significant share of numerous commercial bio-formulations of *T. harzianum*. The success of biological control of phytopathogens using *Trichoderma* does not rely solely on effective antagonists but also on the mode of delivery or application on the seed, root, and soil (Mathre et al., 1999). Timing of delivery and application is also crucial, and *Trichoderma* is usually only effective as a preventative measure but can be integrated with other disease management options especially when the disease has already been established.

Conclusion

Plant diseases can be controlled using chemical pesticides/insecticides. But many phytopathogens have gained resistance against these agrochemicals. Most of the chemicals can be hardly eliminated from the environment and it adversely affects soil microbial functions, soil fertility, and the growth of plants. These agrochemicals are also harmful to humans, animals, and the environment at large. With increased public concern to minimize the use of high-cost pesticides and insecticides in agriculture, an alternative bio-control strategy is the need of the hour. Target-specific fungal BCAs are quite adequate for the management of various crop diseases, phytopathogens, plant-parasitic nematodes, and pests. *Trichoderma* is an important natural biocontrol agent against a wide range of fungal pathogens. This chapter provides an overview of the significance of *Trichoderma* in agriculture, their mechanisms of action, current research on how *Trichoderma* are used, and how this research can improve future agriculture practices.

References

Ahmed A S, Ezziyyani M, Sanchez, C P, Candela M E. Effect of chitin on biological control activity of *Bacillus* spp. and *Trichoderma harzianum* against root rot disease in pepper (*Capsicum annuum*) plants. *Eur. J. Plant. Pathol.* (2003) 109: 633-637.

Almassi F, Ghisalberti, E L, Narbey, M J, Sivasithamparam K. New antibiotics from strains of *Trichoderma harzianum*. *J. Nat. Prod.* (1991) 54: 396-402.

Amin F, Razdan V K. Potential of *Trichoderma* species as biocontrol agents of soil borne fungal propagules. *J. Phytol.* (2010) 2.

Bastakoti S, Belbase S, Manandhar S, Arjyal C. *Trichoderma* species as biocontrol agent against soil borne fungal pathogens. *Nepal. J. Biotechnol.* (2017) 5: 39-45.

Bharti M K, Sharma A K, Pandey A K, Mall R. Physiological and biochemical basis of growth suppressive and growth promotory effect of *Trichoderma* strains on tomato plants. *Natl. Acad. Sci. Lett.* (2012) 35: 355-359.

Brotman Y, Kapuganti G J, Viterbo A *Trichoderma*. *Curr. Biol.* (2010) 20: 390-391.

Chaverri P, Branco-Rocha F, Jaklitsch W, Gazis R, Degenkolb T, Samuels, G J. Systematics of the *Trichoderma harzianum* species complex and the re-identification of commercial biocontrol strains. *Mycologia* (2015) 107: 558-590.

Chet I. *Trichoderma*-Application, mode of action, and potential as a biocontrol agent of soil-born pathogenetic fungi. In: *Innovative approaches to plant disease control. Chet I (Ed.).* John Wiley and Sons (1987) 137-160.

Das M M, Haridas M, Sabu A. Biological control of black pepper and ginger pathogens, *Fusarium oxysporum*, *Rhizoctonia solani* and *Phytophthora capsici*, using *Trichoderma* spp. *Biocatal, Agric, Biotechnol,* (2019) 17: 177-183.

Das M M, Abdulhameed S. Agro-processing residues for the production of fungal bio-control agents. In *Valorisation of Agro-industrial Residues–Volume II: Non-Biological* (2020) 107-126, Springer, Cham.

De Marco J L, Lima L H C, de Sousa M V, Felix C R. A *Trichoderma harzianum* chitinase destroys the cell wall of the phytopathogen *Crinipellis perniciosa*, the causal agent of witches' broom disease of cocoa. *World J. Microbiol. Biotechnol.* (2000) 16: 383-386.

El Komy M H, Saleh A A, Eranthodi A, Molan Y Y. Characterization of Novel *Trichoderma asperellum* isolates to select effective biocontrol agents against tomato *Fusarium wilt*. *Plant Pathol. J.* (2015) 31: 50-60.

Elad Y, Chet J, Katan J. *Trichoderma harzianum*: A biocontrol effective against *Sclerotium rolfsii* and *Rhizoctonia solani*. *J. Phytopathol.* (1980) 70: 119-121.

El-Katatny M H, Somitsch W, Robra K H, El-Katatny M S, Gübitz G M. Production of chitinase and 1, 3-glucanase by *Trichoderma harzianum* for control of the phytopathogenic fungus *Sclerotium rolfsii*. *Food Technol. Biotechnol.* (2000) 38: 173-180.

Hhmau H, Wijesundera, R L C, Chandrasekharan N V, Wijesundera W S S, Kathriarachchi H S. Isolation and characterization of *Trichoderma erinaceum* for antagonistic activity against plant pathogenic fungi. *Curr. Res. Environ. Appl. Mycol.* (2015) 5: 120-127.

Howell C R. Mechanisms employed by *Trichoderma* species in the biological control of plant diseases: the history and evolution of current concepts. *Plant Dis.* (2003) 87: 4-10.

Hukma R, Pandey R N. Efficacy of bio-control agents and fungicides in the management of wilt of pigeon pea. *Indian Phytopathol.* (2011) 64: 269-271.

Jyoti S, Singh D. Fungi as biocontrol agents in sustainable agriculture. Microbes and environmental management. Studium Press, India (2016) 172-194.

Kapoor A S. Biocontrol potential of *Trichoderma* spp. against important soil-borne diseases of vegetable crops. *Indian Phytopathol.* (2011).

Khan A, Williams K L, Nevalainen H K. 2004. Effects of *Paecilomyces lilacinus* protease and chitinase on the eggshell structures and hatching of *Meloidogyne javanica* juveniles. *Biol. Control* (2004) 31: 346-352.

Khan A, Williams K L, Nevalainen H K M. Control of plant-parasitic nematodes by *Paecilomyces lilacinus* and *Monacrosporium lysipagum* in pot trials. *BioControl* (2006) 51: 643-658.

Khan R A A, Najeeb S, Mao Z, Ling J, Yang Y, Li Y, Xie B. Bioactive secondary metabolites from *Trichoderma* spp. against phytopathogenic bacteria and root-knot nematode. *Microorganisms* (2020) 8: 401.

Kumar N, Kumar K, Seshakiran K. Management of *Phytophthora* foot rot disease in black pepper. *Green Farming* (2012) 3: 583-585.

Kumar S. *Trichoderma*: A biological weapon for managing plant diseases and promoting sustainability. *Int. J. Agric. Sci. Med. Vet.* (2013) 1: 106-121.

Kumar S, Manibhushan T, Archana R. *Trichoodema*: Mass production, formulation, quality control, delivery and its scope in commercialization in India for the management of plant diseases. *Afr. J. Agric. Res.* (2014) 9: 3838-3852.

Kumar V, Sharma D D, Babu A M, Datta R K. SEM studies on the hyphal interaction between a biocontrol agent *T. harzianum* and mycopathogen *Fusarium* causing root rot disease in mulberry. *Ind. J. Seric* (1998) 37:17-20.

La Spada F, Stracquadanio C, Riolo M, Pane A, Cacciola S O. *Trichoderma* counteracts the challenge of *Phytophthora nicotianae* infections on tomato by modulating plant defense mechanisms and the expression of crinkler, necrosis-inducing *Phytophthora* protein 1, and cellulose-binding elicitor lectin pathogenic effectors. *Front. Plant Sci.* (2020) 11, p.583539.

Li G Q, Huang H C, Acharya S N, Erickson R S. Effectiveness of *Coniothyrium minitans* and *Trichoderma atroviride* in suppression of *Sclerotinia* blossom blight of alfalfa. *Plant Pathol.* (2005) 54: 204-211.

Mascarin G M, Bonfim Junior M F, de Araújo Filho J V. *Trichoderma harzianum* reduces population of *Meloidogyne incognita* in cucumber plants under greenhouse conditions. *Embrapa Arroz e Feijão-Artigo em periódico indexado (ALICE)* (2012).

Matarese F, Sarrocco S, Gruber S, Seidl-Seiboth, V, Vannacci G. Biocontrol of *Fusarium* head blight: interactions between *Trichoderma* and mycotoxigenic *Fusarium. Microbiology* (2012) 158:98-106.

Mathre D E, Cook R J, Callan N W. From discovery to use traversing the world of commercializing biocontrol agents for plant disease control. *Plant Dis* 83: 972-983.

Monte, E., 2001. Understanding *Trichoderma*: Between agricultural biotechnology and microbial ecology. *Int. Microbiol.* (1999) 4:1-4.

Naher L, Yusuf U K, Tan S G, Siddiquee S, Islam M R. *In vitro* and *in vivo* biocontrol performance of *Trichoderma harzianum* rifai on *Ganoderma boninense* Pat. related to pathogenicity on oil palm (*Elaeis guineensis* Jacq.) *J. Pure Appl.* (2014) 8:973-978.

Nakkeeran S, Krishnamoorthy A S, Ramamoorthy V, Renukadevi S. *Microbial inoculants in plant disease control. J. Ecobiol.* (2002) 14:83-94.

Naserinasab F, Sahebani N and Etebarian H R. Biological control of *Meloidogyne javanica* by *Trichoderma harzianum* BI and salicylic acid on tomato. *Afr. J. Food Sci.* (2011) 5: 276-280.

Nusaibah S A, Musa H. Review report on the mechanism of *Trichoderma* spp. as biological control agent of the basal stem rot (BSR) disease of *Elaeis guineensis*. In *Trichoderma-The most widely used fungicide* (2019) 79-90.

Pan S, Das A. Control of cowpea (*Vigna sinensis*) root and collar rot (*Rhizoctonia solani*) with some organic formulations of *Trichoderma harzianum* under field condition. *J. Plant Prot. Sci.* (2011) 3: 20-25.

Parmar H J, Hassan M M, Bodar N P, Umrania V V, Patel S V, Lakhani H N. *In vitro* antagonism between phytopathogenic fungi *Sclerotium rolfsii* and *Trichoderma* strains. *Int. J. Appl. Sci. Biotechnol.* (2015) 3: 16-19.

Patil A S, Chakranarayan M R. Production and characterization of *Trichoderma* metabolites: a new approach for selective bioremediation. *Recent trends in PGPR research for sustainable crop productivity* (2016) p.29.

Paulitz T C, Ahmad J S, Baker R. Integration of *Pythium nunn* and *Trichoderma harzianum* isolate T-95 for the biological control of *Pythium* damping-off of cucumber. *Plant Soil* (1990) 121: 243-250.

Puyam A. Advent of *Trichoderma* as a bio-control agent-a review. *J. Appl. Nat. Sci.* (2016) 8: 1100-1109.

Samuels GJ. *Trichoderma*: a review of biology and systematic of the genus. *Mycol Res* (1996) 100: 923-935.

Sandhu S S, Sharma A K, Beniwal V, Goel G, Batra P, Kumar A, Jaglan S, Sharma A K, Malhotra S. Myco-biocontrol of Insect pests: Factors involved, mechanism, and regulation. *J. pathog.* (2012).

Sariah M, Choo C W, Zakaria H, Norihan, M S. Quantification and characterization of *Trichoderma* spp. from different ecosystems. *Mycopathologia* (2005) 159: 113-117.

Schrank A, Vainstein M. *Metarhizium anisopliae* enzymes and toxins. *Toxicon* (2010) 56: 1267-1274.

Shah M M, Afiya H. Introductory Chapter: Identification and Isolation of *Trichoderma* spp.-Their significance in agriculture, human health, industrial and environmental application. In *Trichoderma-The most widely used fungicide. IntechOpen* (2019).

Shakeri J, Foster H A. Proteolytic activity and antibiotic production by *Trichoderma harzianum* in relation to pathogenicity to insects. *Enzyme Microb. Technol.* (2007) 40: 961-968.

Sharma A, Diwevidi V D, Singh S, Pawar K K, Jerman M, Singh B, Singh, S, Srivastawa. Biological control and its important in agriculture. *Int. J. Biotechnol. Bioeng. Res.* (2013) 4: 175-180.

Sharma K K. *Trichoderma* in Agriculture: An overview of global scenario on research and its application. *Int. J. Curr. Microbiol. Appl. Sci.* (2018) 7: 1922-1933.

Shoresh M, Harman G E, Mastouri F. Induced systemic resistance and plant responses to fungal biocontrol agents. *Annu. Rev. Phytopathol.* (2010) 48: 21-43.

Singh S, Kumar R, Yadav S, Kumari P, Singh R K and Kumar CR. Effect of bio-control agents on soil borne pathogens: A review. *J. Pharmacogn. Phytochem.* (2018) 7: 406-411.

Sivan, A, Chet I. The possible role of competition between *Trichoderma harzianum* and *Fusarium oxysporum* on rhizosphere colonization. *Phytopathol.* (1989) 79: 198-203.

Strakowska J, Błaszczyk L, Chełkowski J. The significance of cellulolytic enzymes produced by *Trichoderma* in opportunistic lifestyle of this fungus. *J. Basic Micro* (2014) 54: S2-S13.

Tjamos E C, Papavizas G C and Cook R J eds. *Biological control of plant diseases: progress and challenges for the future* (Vol. 230). Springer Science & Business Media (2013).

Uddi M N, Rahman U, Uddin N, Muhammad M. Effect of *Trichoderma harzianum* on tomato plant growth and its antagonistic activity against *Phythium ultimum* and *Phytopthora capsici*. *Egyptian J. Biol. Pest Co* (2018) 28: 1-6.

Usta C. Microorganisms in biological pest control-a review (bacterial toxin application and effect of environmental factors). In *Current progress in biological research. IntechOpen* (2013) 287-317.

Vinale F, Flematti G, Sivasithamparam K, Lorito M, Marra R, Skelton B W, Ghisalberti E L. Harzianic acid, an antifungal and plant growth promoting metabolite from *Trichoderma harzianum*. *J. Nat. Prod.* (2009) 72: 2032-2035.

Vinale, F, Marra R, Scala F, Ghisalberti E L, Lorito, M, Sivasithamparam K. Major secondary metabolites produced by two commercial *Trichoderma* strains active against different phytopathogens. *Lett. Appl. Microbiol.* (2006) 43: 143-148.

Vinale F, Nigro M, Sivasithamparam K, Flematti G, Ghisalberti E L, Ruocco M, Varlese R, Marra R, Lanzuise S, Eid A, Woo S L. Harzianic acid: a novel siderophore from *Trichoderma harzianum*. *FEMS Microbiol. Lett.* (2013) 347:123-129.

Vizcaino J A, Sanz L, Cardoza R E, Monte E, Gutierrez S. Detection of putative peptide synthetase genes in *Trichoderma* species. Application of this method to the cloning of a gene from *T. harzianum* CECT 2413. *FEMS Microbiol. Let.* (2005) 244: 139-148.

Yao, Y, Yan L, Zhiqun C, Baiqin Z, Litian Z, Ben N, Junli M, Aijun L, Jianmin Z, Qi W. Biological control of potato late blight using isolates of *Trichoderma*. *Amer. J. Potato Res.* (2016) 93: 33-42.

Yashaswini C, Sudarsanam V K. *Entomopathogenic fungi as biological controller* (2017).

Zeilinger S, Gruber S, Bansal R, Mukherjee P K. Secondary metabolism in *Trichoderma* chemistry meets genomics. *Fungal Biol. Rev.* (2016) 30: 74-90.

Zin N A, Badaluddin N A. Biological functions of *Trichoderma* spp. for agriculture applications. *Ann. Agric. Sci.* (2020) 65: 168-178.

Chapter 5

Trichoderma Uses in Agriculture: A Multipurpose Tool for Biological Control and Plant Growth

Francisco Wilson Reichert Junior[1,*]
José Luís Trevizan Chiomento[1]
Crislaine Sartori Suzana-Milan[1]
Brenda Tortelli[1]
Aline Frumi Camargo[2]
Jéssica Mulinari[3]
Altemir José Mossi[2]
and Helen Treichel[2]
[1]Passo Fundo University, Passo Fundo, Brazil
[2]Federal University of Fronteira Sul, Erechim, Brazil
[3]Federal University of Santa Catarina, Florianópolis, Brazil

Abstract

Trichoderma is a fungus widely used in agriculture due to its beneficial effects on plant growth and disease control. This fungus can colonize the plant root system and provide them with nutrients such as nitrogen, phosphorus, and potassium. Moreover, *Trichoderma* can also produce phytohormones and phytoregulators, which promote plant growth and development. In addition, *Trichoderma* is known for its antagonistic activity against several plant pathogens, making it a potential biocontrol

---

* Corresponding Author's Email: helentreichel@gmail.com.

In: Trichoderma: Taxonomy, Biodiversity and Applications
Editor: Michael S. Mouton
ISBN: 979-8-88697-946-6

agent for plant disease control. *Trichoderma* is usually considered a biocontrol agent against fungal pathogens, but it has also demonstrated its potential as a bioherbicide and bioinsecticide. It can inhibit the growth and development of a variety of weed species. *Trichoderma*'s action relies on the production of enzymes that break down the weed's cell walls, leading to its death. In addition, these fungi are effective against a wide range of pests, including nematodes and insects. *Trichoderma* is a bioinsecticide because it can produce compounds such as lectins, proteases, and secondary metabolites with insecticidal properties. Overall, the use of *Trichoderma* in agriculture has the potential to increase crop yields, reduce pollution, and promote sustainable agriculture.

Keywords: *Trichoderma*, biological control, sustainable agriculture

Introduction

The search for ecologically balanced agriculture is one of the main challenges of the 21st century. The tremendous increase in agricultural production due to the green revolution helped increase productivity; however, it led to several problems, such as environmental pollution and the use of products toxic to the environment and humans (Raney, 2009). Furthermore, the repeated use of several active herbicides, fungicides, and insecticide ingredients has led to the emergence of resistant species, which are increasingly difficult to manage. Therefore, searching for more ecologically correct alternatives to control agricultural pests and diseases is of fundamental importance (Galon et al., 2016).

Biological control of pests and diseases has advanced a lot in recent years. It uses living organisms, or their by-products, to control agricultural pests and diseases. Fungi are among the most studied organisms for biological control. Different genera of fungi can be used for various purposes; however, around 90% of biocontrol agents for agricultural diseases belong to other strains of *Trichoderma* (Köhl et al., 2019; Hermosa et al., 2012). *Trichoderma*'s main control methods against phytopathogenic fungi are mycoparasitism, competition, and antibiosis (Ahluwalia et al., 2015).

In addition to the potential to control phytopathogenic fungi, *Trichoderma* has already been reported as a potential biocontrol of insects. *Trichoderma* acts directly as an entomopathogen through parasitism and the production of secondary metabolites with insecticidal effects, anti-food compounds, and repellent components. It can also work indirectly by activating the plant's defense system, attracting natural enemies or parasitic insects (Poveda, 2021).

Bioherbicidal effects have already been investigated for fungi of this genus. Reichert Junior et al. (2019) observed the phytotoxic effect of *Trichoderma koningiopsis* on *Euphorbia heterophylla*, and the main phytotoxic methods were enzymatic activity and phytotoxin production during the fermentation process. Also, Zhu et al. (2020) observed high levels of mortality of *Elsholtzia densa* Benth., *Polygonum lapathifolium* L., *Lepyrodiclis holosteoides* (C.A. Mey.) Fenzl ex Fisch. & C.A. Mey., *Avena fatua* L., *Chenopodium album* L., and *Polygonum aviculare* when exposed to a bioherbicide based on *Trichoderma polysporum*.

Besides the potential for biocontrol of agricultural pests, fungi of the genus *Trichoderma* can be used as plant growth promoters. They can stimulate plant growth by producing phytohormones and phytostimulants and increasing the roots' potential for nutrient searching (Zhao et al., 2014; Jaroszuk-Ściseł et al., 2019). Several species positively responded to *Trichoderma* inoculation, indicating a relatively generalist symbiosis (Harman et al., 2004).

Therefore, the genus *Trichoderma* presents itself as a promising alternative for the biological control of several agricultural pests and diseases, in addition to the biocontrol of weeds and plant growth promotion. This multifunctional microorganism is a fundamental tool in the search for a more ecologically balanced agriculture, something essential in the current moment of our planet.

Plant Growth Promoter

Fungi belonging to the genus *Trichoderma* are identified as the primary critical components of plant biostimulants (Woo et al., 2022). After a cascade of signals between the plant host and the *Trichoderma* spp., followed by the establishment of symbiosis, this rhizofungus has the potential to improve the growth and development of its plant partner. The *Trichoderma*-plant interface, resulting from root colonization and the multitude of compounds produced by the microorganism, activates biochemical and genetic pathways linked to plant defense responses to biotic and abiotic stresses (Harman et al., 2004). Thus, synthesizing phytohormones and phytoregulators is the main stimulating factor in almost all plant growth and development (Jaroszuk-Ściseł et al., 2019).

Trichoderma spp. potentiates the capacity of the roots to explore the growth medium (soil and substrate), which contributes positively to the availability and absorption of nutrients. Nutrient uptake is improved due to the conversion of nutrients from their unavailable forms to their available formats. For example, *Trichoderma* strains produce and release coumaric, glucuronic, and citric acids, which aid in supplying phosphorus ions to plants (Zhao et al., 2014). Inoculating sugarcane (*Saccharum officinarum* L.) with *T. viride* improved the absorption of the macronutrients nitrogen, potassium, and phosphorus (Yadav et al., 2009). Root and shoot growth of tomato (*Solanum lycopersicum* L.) cultivated with *T. harzianum* increased the absorption of micronutrients such as copper and zinc (Li et al., 2015).

In addition to the beneficial effects related to the acquisition and absorption of nutrients, *Trichoderma* spp. intensifies the production of phytohormones in the plant host, such as indole-3-acetic acid and gibberellic acid (Bader et al., 2020), both responsible for plant elongation. Thus, the literature reports that this rhizofungus improves plant morphology by promoting increased biomass, height, and architecture of plant hosts' shoot and root systems (Halifu et al., 2019). In cucumber (*Cucumis sativus* L.), the use of *T. harzianum* potentiated the root biomass (Yedidia et al., 2001), and in cabbage (*Brassica oleracea* var. *capitata* L.), the species *T. longipile* and *T. tomentosum* increased the leaf area and the seedling biomass (Rabeendran et al., 2000).

The promotion of plant growth also occurs because *Trichoderma* spp. regulates physiological processes in plants, such as photosynthesis, gas exchange, stomatal conductance, and water use efficiency (Sood et al., 2020). Rice plants (*Oryza sativa* L.) treated with *Trichoderma* sp. SL2 had stimuli in photosynthetic rate, stomatal conductance, and water use efficiency (Doni et al., 2014). In tomatoes, the use of *T. harzianum* increased the levels of chlorophyll in the leaves (Vukelić et al., 2021). The literature also shows that *Trichoderma* spp. promotes plant growth in stressful environments. Tomato plants grown under thermal stress had the effect of cold mitigated on plants inoculated with *T. harzianum* AK20G (Ghorbanpour et al., 2018). Using *T. longibrachiatum* T6 in wheat (*Triticum aestivum* L.) grown in a saline environment intensified the gene expression of superoxide dismutase, catalase, and peroxidase in plants (Zhang et al., 2015).

All these synergistic effects promoted by *Trichoderma* spp. provide a less stressful environment for plant hosts. Therefore, using this microorganism contributes to achieving optimal productivity and quality of crops established in fields and orchards. Harzianic acid and 6-pentyl-α-pyrone produced by *T.*

harzianum and *T. atroviride* increased grapevine yield and quality (*Vitis vinifera* L.) (Pascale et al., 2017). Applications of *T. harzianum* T22 and TH1, and *T. virens* GV41 increased strawberry production (*Fragaria* X *ananassa* Duch.) and favored the accumulation of anthocyanins in ripe fruits (Lombardi et al., 2020).

Biofungicide

The genus Trichoderma has been successfully used to control several plant diseases due to its ability to reduce survival, growth, or infections caused by phytopathogens. These different abilities result in several studies of mechanisms for its management, in which the biocontrol agent antagonizes the phytopathogen, resulting in various types of interactions between the organisms (Pal; Gardener, 2006).

Mycoparasitism is one possible interaction between the fungus and the phytopathogen. In this case, the microorganism obtains nutrients from the host's living and functioning cells. Sequential events are required for this mechanism, including host recognition, attack, penetration, and death. It starts with the chemical stimulus of fungi, which attracts the antagonist and induces chemotropic responses. Next, there is the recognition between the antagonist and the phytopathogen. Then the antagonist grows, winding itself along the host's hyphae and secreting different lytic enzymes, such as chitinase, glucanase, and pectinase, which degrade the cell wall (Benítez et al., 2004). The best-known example of mycoparasitism is the ability of different *Trichoderma* isolates to colonize sclerotia, survival structures of the phytopathogen *Sclerotinia sclerotiorum*, the causal agent of white mold (Smith et al., 2013).

Trichoderma spp. can also decompose phytopathogenic fungi without any physical contact, producing antimicrobial compounds capable of inhibiting the growth of the phytopathogen, a process called antibiosis (Wu et al., 2014). Antibiosis occurs during the interaction process of *Trichoderma* strains, which produce volatile or non-volatile toxic metabolic compounds, suppressing the colonization of the affected organism (Reino et al., 2008). The *in vitro* antifungal activity of many secondary metabolites produced by *Trichoderma* against *Botrytis*, *Fusarium*, *Rhizoctonia*, *Sclerotinia*, *Colletotrichum*, *Phytophthora*, and *Pythium* was already reported (Hermosa et al., 2012). *T. harzianum* and *T. virens* are the most effective biocontrol agents concerning the antimicrobial compounds they produce: gliovirin and pyrone, respectively (Ghazanfar et al., 2018).

Competition has been considered one of Trichoderma's most efficient mechanisms of action. It is related to the ability to compete for the same nutrient or space with other organisms, reducing the amount available to others. *Trichoderma* is regarded as an aggressive competitor, mainly to soil dwellers, since it proliferates rapidly and suppresses the growth of the phytopathogen population in the rhizosphere (Cuervo-Parra et al., 2014). *Trichoderma* species have a remarkable ability to mobilize and absorb nutrients from the soil compared to other microorganisms (Verma et al., 2007), besides being more resistant to different toxic compounds (Benítez et al., 2004). *Rhizoctonia* and *Pythium* are phytopathogens controlled by other *Trichoderma* species in the soil through competition (Ruocco et al., 2009).

Bioinsecticide

The genus *Trichoderma* is composed of species beneficial to plants, with importance in agriculture as a biological control agent (Monte, 2023). Its entomopathogenic action has already been described as a function of gliotoxin, a secondary metabolite produced by *Trichoderma* toxic to animal cells and known mainly for being involved in the antagonism of microorganisms in the rhizosphere (Vargas et al., 2014).

The biocontrol of insect pests mediated by *Trichoderma* is described as a function of the fungus-plant-animal-bacteria multitrophic interaction, which leads to modifying the bacterial microbiome in phytophagous larvae that feed on the leaves of a plant whose metabolome was modified after colonization by *Trichoderma*. In addition, *Trichoderma* can also reprogram the plant's immune memory by activating the mechanism known as priming, which leads to a quick and robust plant defense response that is longer-lasting and heritable (Morán-Diez et al., 2021).

Trichoderma can colonize and live endophytically in plant roots, which allows different strains to communicate with the plants, cause their growth and induce priming. Concerning herbivores, *Trichoderma* is interested in maintaining the quality of the host plant as a function of symbiosis. In this way, it developed the ability to activate plant responses, such as 1) positively regulating genes involved in oxidative explosion reactions, 2) producing alterations in the metabolome that result in repellent, anti-nutritive and toxic effects on herbivores or influencing the production of proteins in the insect's intestine; 3) increase the expression of genes that encode protective enzymes (Woo et al., 2022). *Trichoderma* spp. can activate one or more of these mechanisms, increasing the plant's defense against insect pests (Monte, 2023).

The direct action of *Trichoderma* on insect pests of the order *Lepidoptera* occurs through rupture and permeability of the peritrophic membrane of the midgut. Tomato plants colonized by *T. afroharzianum* negatively influence the development and survival of the pest *Spodoptera littoralis* by altering the intestinal microbiota of larvae and their nutritional support to the host (Di Lelio et al., 2021). In addition, the literature confirms the action of *Trichoderma* on aphids (Coppola et al., 2019), thrips (Muvea et al., 2014), silverleaf whitefly (Jafarbeigi et al., 2020), beetles (Praprotnik et al., 2021) and leafhopper vector of *Xylella fastidiosa* (Ganassi et al., 2023). *Trichoderma* can also exert biocontrol by releasing volatile compounds, which attract parasitoids and insect pest predators, or have antixenosis or non-preference action, which can reduce insect attack (Coppola et al., 2019), characteristics that *Trichoderma* spp. acquired during their evolutionary leaps (Woo et al., 2022).

Bioherbicide

Fungi are among the most studied organisms for the development of bioherbicides. In addition to the primary metabolites necessary for their growth and functioning, these organisms produce secondary metabolites which are not fundamental to their metabolism (Bell, 1981). These metabolites are produced during infectious processes or stress. When these substances are toxic to plants, they are called phytotoxins (Klaic et al., 2015; Galon et al., 2016). These phytotoxins can be a source of bioherbicides and can be extracted from culture media or fermentation processes of fungi. Several fungi have already been identified with bioherbicide potential, including fungi of the genus *Fusarium*, *Colletotrichum*, *Aspergillus*, and *Penicillium* (Tremacoldi et al., 2006).

Fungi of the genus *Trichoderma* are widely studied as biocontrol of phytopathogenic fungi and insect pests (Poveda, 2021). However, studies have shown that these fungi can also have a bioherbicidal effect, causing a phytotoxic effect against dicotyledonous plants (Reichert Junior et al., 2019). The bioherbicide mode of action may be related to the activity of enzymes that can facilitate the entry of phytotoxic substances, degrading the wax layer on the leaf surface and the plant cell wall. Among the main enzymes involved in this process are lipases, peroxidases, cellulases, and amylases (Reichert Junior et al., 2019; Urlich et al., 2021).

Trichoderma harzianum, *Trichoderma virens*, *Trichoderma reesei*, *Trichoderma pseudokoningii*, *Trichoderma viridae*, and *Trichoderma koningiopsis* are species that have already shown herbicidal potential in different studies (Heraux et al., 2005; Javaid and Ali, 2011; Reichert Junior et al., 2019). Although many *Trichoderma* species are being studied to develop bioherbicides, no product based on this microorganism is available. The main limiting factor for using fungi as bioherbicides are their specificity, specific environmental conditions for good efficiency, and possible effects on non-target organisms (Kremer, 2005; Rush et al., 2021). However, the search for fungi-based bioherbicides finds in the genus *Trichoderma* one of the best alternatives for developing biological products for weed control. Zhu et al. (2020) observed high levels of mortality in *Elsholtzia densa* Benth., *Polygonum lapathifolium* L., *Lepyrodiclis holosteoides* (C.A. Mey.) Fenzl ex Fisch. & C.A. Mey., *Avena fatua* L., *Chenopodium album* L., and *Polygonum aviculare* when exposed to a bioherbicide based on *Trichoderma polysporum*, with virulence being associated with the combination of conidia and metabolites produced during fermentation.

Another alternative is the use of bioherbicides associated with adjuvants and commercial herbicides. The benefit of this type of interaction is the possibility of controlling weeds with a reduction in the dose of chemical products. Camargo et al. (2019) observed that the mixture of fungal extracts based on *Trichoderma koningiopsis* with synthetic herbicides achieved weed control with only half the recommended dose of these herbicides, yet weeds resistant to active principles such as glyphosate were controlled by this herbicide when mixed with fungal extracts. In this sense, fungi of the genus *Trichoderma* are a promising alternative for developing biological products for weed control.

Conclusion

Research involving the development of biological products has advanced a lot in recent years. It should be encouraged since current agriculture faces several challenges, sustainability being one of the main ones. Therefore, *Trichoderma* proves to be a surprising ally in the search for sustainable agriculture. Given the multifunctionality of fungi of the genus *Trichoderma* for agriculture, this microorganism becomes an essential tool in the search for more ecologically sustainable products.

References

Ahluwalia, V.; Kumar, J.; Rana, V. S.; Sati, O. P.; Walia, S. (2015). Comparative evaluation of two *Trichoderma harzianum*strains for major secondary metabolite production and antifungal activity. *Nat. Prod. Res*, *29*, 914–920.

Bader, A. N., Salerno, G. L., Covacevich, F. and Consolo, V. F. (2020). Native *Trichoderma harzianum* strains from Argentina produce indole-3 acetic acid and phosphorus solubilization, promote growth and control wilt disease on tomato (*Solanum lycopersicum* L.). *Journal of King Saud University – Science,* 32(1):867–873. doi: 10.1016/j.jksus.2019.04.002.

Bell, A. A. (1981). Biochemical mechanisms of disease resistance. *Ann. Rev. Plant Physiol.* 32:21-81.

Benítez, T., Rincón, A. M., Limón, M. C. and Codón, A. C. (2004). Biocontrol mechanisms of *Trichoderma strains*. *Internacional Microbiology,* 4:249–260.

Camargo, A. F., Stefanski, F. S., Scapini, T., Weirich, S. N., Ulkovski, C., Carezia, C., Bordin, E. R., Rossetto, V., Júnior, F. R., Galon, L., Fongaro, G., Mossi, A. J., & Treichel, H. (2019). Resistant weeds were controlled by the combined use of herbicides and bioherbicides. *Environmental Quality Management,* 1-6. doi: 10.1002/tqem.21643.

Coppola, M., Cascone, P., Di Lelio, I., Woo, S. L., Lorito, M., Rao, R., Pennacchio, F., Guerrieri, E. and Digilio, M. C. (2019). *Trichoderma atroviride* P1 colonization of tomato plants enhances both direct and indirect defense barriers against insects. *Frontiers in Physiology,* 10:813. doi: 10.3389/fphys.2019.00813.

Cuervo-Parra, J. A., Snchez-Lpez, V., Romero-Cortes, T. and Ramrez-Lepe, M. (2014). Hypocrea/*Trichoderma viridescens* ITV43 with potential for biocontrol of *Moniliophthora roreri* Cif Par, *Phytophthora megasperma* and *Phytophthora capsici*. *African Journal of Microbiology Research,* 8:1704–1712. doi: 10.5897/AJMR2013. 6279.

Di Lelio, I., Coppola, M., Comite, E., Molisso, D., Lorito, M., Woo, S. L., Pennacchio, F., Rao, R. and Digilio, M. C. (2021). Temperature differentially influences the capacity of *Trichoderma* species to induce plant defense responses in tomato against insect pests. *Frontiers in Plant Science,* 12:678830. doi: 10.3389/fpls.2021.678830.

Doni, F., Isahak, A., Radziah, C., Zain, C. M., Ariffin, S. M., Mohamad, W. N. W. and Yusoff, W. M. W. (2014). Formulation of *Trichoderma* sp. SL2 inoculants using different carriers for soil treatment in rice seedling growth. SpringerPlus, 3:532. doi: 10.1186/2193-1801-3-532.

Galon, L., Mossi, A. J., Reichert Junior, F. W., Reik, G. G., Treichel, H., Forte, C. T. (2016). Biological Weed management – A short review. *Revista Brasileira de Herbicidas,* 15(1): 116-125. doi: http://dx.doi.org/10.7824/rbh.v15i1.452.

Ganassi, S., Di Domenico, C., Altomare, C., Samuels, G. J., Grazioso, P., Di Cillo, P., Pietrantonio, L. and De Cristofaro, A. (2023). Potential of fungi of the genus *Trichoderma* for biocontrol of *Philaenus spumarius*, the insect vector for the quarantine bacterium *Xylella fastidosa*. *Pest Management Science,* 79(2):719–728. doi: 10.1002/ps.7240.

Ghazanfar, M. U., Raza, M., Raza, W. and Qamar, M. I. (2018). *Trichoderma* as potential biocontrol agent, its exploitation in agriculture: a review. *Plant Protection,* 03:109–135.

Ghorbanpour, A., Salimi, A., Ghanbary, M. A. T., Pirdashti, H. and Dehestani, A. (2018). The effect of *Trichoderma harzianum* in mitigating low temperature stress in tomato (*Solanum lycopersicum* L.) plants. *Scientia Horticulturae,* 230:134–141. doi: 10.1016/j.scienta.2017.11.028.

Halifu, S., Deng, X., Song, X. and Song, R. (2019). Effects of two *Trichoderma* strains on plant growth, rhizosphere soil nutrients, and fungal community of *Pinus sylvestris* var. mongolica annual seedlings. *Forests,* 10:758. doi: 10.3390/f10090758.

Harman, G. E., Howell, C. R., Viterbo, A., Chet, I. and Lorito, M. (2004). *Trichoderma* species — opportunistic, avirulent plant symbionts. *Nature Reviews Microbiology,* 2:43–56. doi: 10.1038/nrmicro797.

Heraux, F. M. G., Hallett, S. G., Ragothama, K. G., Weller, S. C., (2005). Composted chicken manure as a medium for the production and delivery of *Trichoderma virens* for weed control. *Hortscience* 40, 1394–1397.

Hermosa, R., Viterbo, A., Chef, I. and Monte, E. (2012). Plant-beneficial effects of *Trichoderma* and of its genes. *Microbiology,* 158:17–25. doi: 10.1099/mic.0.052274-0.

Jafarbeigi, F., Samih, M. A., Alaeli, H. and Shirani, H. (2020). Induced tomato resistance against *Bemisia tabaci* triggered by salicylic acid, β-aminobutyric acid, and *Trichoderma*. *Neotropical Entomology,* 49:456–467. doi: 10.1007/s13744-020-00771-0.

Jaroszuk-Ściseł, J., Tyśkiewicz, R., Nowak, A., Ozimek, E., Majewska, M., Hanaka, A., Tyśkiewicz, K., Pawlik, A. and Janusz, G. (2019). Phytohormones (auxin, gibberellin) and ACC deaminase in vitro synthesized by the mycoparasitic *Trichoderma* DEMTkZ3A0 strain and changes in the level of auxin and plant resistance markers in wheat seedlings inoculated with this strain conidia. *International Journal of Molecular Sciences,* 20:4923. doi: 10.3390/ijms20194923.

Javaid, A., Ali, S., (2011). Herbicidal activity of culture filtrates of *Trichoderma Spp.* against two problematic weeds of wheat. *Nat. Prod. Res.* 25 (7), 730–740.

Klaic, R., Kuhn, R. C., Foletto, E. L., Dal-Prá, V., Jacques, R. J., Guedes, J. V., Treichel, H., Mossi, A. J., Oliveira, D., Oliveira, J. V., Jahn, S. L., Mazutti, M. A., 2015. An Overview Regarding Bioherbicide and Their Production Methods by Fermentation. *Fungal Biomolecules: Sources, Applications and Recent Developments.* John Wiley & Sons, Ltda,. 183–199. https://doi.org/10.1002/9781118958308.

Köhl, J.; Kolnaar, R.; Ravensberg, W. J. (2019) Mode of action of microbial biological control agents against plant diseases: Relevance beyond efficacy. *Front. Plant Sci.*, 10, 845.

Kremer, R. J. (2005). The role of bioherbicides in weed management. *Biopestic. Int.* 1, 127–141.

Li, R. X., Cai, F., Pang, G., Shen, Q. R., Li, R. and Chen, W. (2015). Solubilisation of phosphate and micronutrients by *Trichoderma harzianum* and its relationship with the promotion of tomato plant growth. *PLoS ONE,* 10(6):0130081. doi: 10.1371/journal.pone.0130081.

Lombardi, N., Caira, S., Troise, A. D., Scaloni, A., Vitaglione, P., Vinale, F., Marra. R., Salzano, A. M., Lorito, M. and Woo, S. L. (2020). *Trichoderma* applications on strawberry plants modulate the physiological processes positively affecting fruit production and quality. *Frontiers in Microbiology,* 11:1364. doi: 10.3389/fmicb.2020.01364.

Monte, E. (2023). The sophisticated evolution of *Trichoderma* to control insect pests. *Proceedings of the National Academy of Sciences,* 120(12):2301971120. doi: 10.1073/pnas.2301971120.

Morán-Diez, M. E., Martínez de Alba, A. E., Rubio, M. B., Hermosa, R. and Monte, E. (2021). *Trichoderma* and the plant heritable priming responses. *Journal of Fungi,* 7:318. doi: 10.3390/jof7040318.

Muvea, A. M., Meyhöfer, R., Subramanian, S., Poehling, H. M., Ekesi, S. and Maniania, N. K. (2014). Colonization of onions by endophytic fungi and their impacts on the biology of *Thrips tabaci*. *PLoS ONE,* 9:108242. doi: 10.1371/journal.pone.0108242.

Pal, K. K. and Gardener, B. M. (2006). Biological control of plant pathogens. *The Plant Health Instructor,* 2:1117–1142. doi:10.1094/PHI-A-2006-1117-02.

Pascale, A., Vinale, F., Manganiello, G., Nigro, M., Lanzuise, S., Ruocco, M., Marra, R., Lombardi, N., Woo, S. L. and Lorito, M. (2017). *Trichoderma* and its secondary metabolites improve yield and quality of grapes. *Crop Protection,* 92:176–181. doi: 10.1016/j.cropro.2016.11.010.

Poveda, J. (2021). Trichoderma as biocontrol agent against pests: new uses for a mycoparasite. *Biol Control* 159:104634. https://doi.org/10.1016/j.biocontrol.2021b.104634.

Praprotnik, E., Lončar, J. and Razinger, J. (2021). Testing virulence of different species of insect associated fungi against yellow mealworm (Coleoptera: Tenebrionidae) and their potential growth stimulation to maize. *Plants,* 10(11):2498. doi: 10.3390/plants10112498.

Rabeendran, N., Moot, D. J., Jones, E. E. and Stewart, A. (2000). Inconsistent growth promotion of cabbage and lettuce from *Trichoderma* isolates. *New Zealand Plant Protection,* 53:143–146. doi: 10.30843/nzpp.2000.53.3667.

Raney, T. (2009) *The State of Food and Agriculture: Livestock in the balance*; Food and Agriculture Organization of the United Nations: Rome, Italy.

Reichert Júnior, F. W., Scariot, M. A., Forte, C. T., Pandolfi, L., Dil, J. M., Weirich, S., Carezia, C., Mulinari, J., Mazutti, M. A., Fongaro, G., Galon, L., Treichel, H., & Mossi, A. J. (2019). New perspectives for weeds control using autochthonous fungi with selective bioherbicide potential. *Heliyon.* doi: https://doi.org/10.1016/j.heliyon.2019.e01676.

Reino, J. L., Guerrero, R. F., Hernández-Galán, R. and Collado, I. G. (2008). Secondary metabolites from species of the biocontrol agent *Trichoderma*. *Phytochemistry Reviews,* 7:89–123. doi: 10.1007/s11101-006-9032-2.

Ruocco, M., Lanzuise, S., Vinale, F., Marra, R., Turrà, D., Woo, S. L. and Lorito, M. (2009). Identification of a new biocontrol gene in *Trichoderma atroviride*: the role of an ABC transporter membrane pump in the interaction with different plant-pathogenic fungi. *Molecular Plant-Microbe Interactions,* 22:291–301. doi: 10.1094/MPMI-22-3-0291.

Rush, T. A., Shrestha, H. K., Gopalakrishnan Meena, M., Spangler, M. K., Ellis, J. C., Labbé, J. L., & Abraham, P. E. (2021) Bioprospecting Trichoderma: a systematic roadmap to screen genomes and natural products for biocontrol applications. *Frontiers in Fungal Biology,* 2, 716511.

Smith, A., Beltrán, C., Kusunoki, M., Cotes, A., Motohashi, K., Kondo, T. and Deguchi, M. (2013). Diversity of soil-dwelling *Trichoderma* in Colombia and their potential as biocontrol agents against the phytopathogenic fungus *Sclerotinia sclerotiorum* (Lib.) de Bary. *Journal of General Plant Pathology,* 79:74–85. doi:10.1007/s10327-012-0419-1.

Sood, M., Kapoor, D., Kumar, V., Sheteiwy, M. S., Ramakrishnan, M., Landi, M., Araniti, F. and Sharma, A. (2020). *Trichoderma*: the "secrets" of a multitalented biocontrol agent. *Plants,* 9(6):762. doi: 10.3390/plants9060762.

Tremacoldi, C. R., Souza Filho, A. P. S., (2006). Toxinas produzidas por fungos fitopatógenos: Possibilidades de uso no controle de plantas daninhas [Toxins produced by phytopathogenic fungi: Possibilities of use in weed control]. *Embrapa*. 22p.

Ulrich, A., Lerin, L. A., Camargo, A. F., Scapini, T., Diering, N. L., Bonafin, F., Gasparetto, I. G., Confortin, T. C., Sansonovicz, P. F., Fabian, R. L., Reichert Júnior, F. W., Treichel, H., Müller, C., & Mossi, A. J. (2021). Alternative bioherbicide based on *Trichoderma koningiopsis*: Enzymatic characterization and its effect on cucumber plants and soil organism. *Biocatalysis and Agricultural Biotechnology.* (36), doi: https://doi.org/10.1016/j.bcab.2021.102127.

Vargas, W. A., Mukherjee, P. K., Laughlin, D., Wiest, A., Morán-Diez, M. E. and Kenerley, C. M. (2014). Role of gliotoxin in the symbiotic and pathogenic interactions of *Trichoderma virens*. *Microbiology,* 160:2319–2330. doi: 10.1099/mic.0.079210-0.

Verma, M., Brar, S. K., Tyagi, R. D., Surampalli, R. Y. and Valéro, J. R. (2007). Antagonistic fungi, *Trichoderma* spp.: panoply of biological control. *Biochemical Engineering Journal,* 37:1–20. doi: 10.1016/j.bej.2007.05.012.

Vukelić, I. D., Prokić, L. T., Racić, G. M., Pešić, M. B., Bojović, M. M., Sierka, E. M., Kalaji, H. M. and Panković, D. M. (2021). Effects of *Trichoderma harzianum* on photosynthetic characteristics and fruit quality of tomato plants. *International Journal of Molecular Sciences,* 22:6961. doi: 10.3390/ijms22136961.

Woo, S. L., Hermosa, R. and Lorito, M. (2022). *Trichoderma*: a multipurpose, plant-beneficial microorganism for eco-sustainable agriculture. *Nature Reviews Microbiology,* 21:312–326. doi: 10.1038/s41579-022-00819-5.

Wu, B., Oesker, V., Wiese, J., Schmaljohann, R. and Imhoff, J. (2014). Two new antibiotic pyridones produced by a marine fungus, *Trichoderma* sp. strain MF106. *Marine Drugs,* 12:1208–1219. doi: 10.3390/md12031208.

Yadav, R. L., Suman, A., Prasad, S. R. and Prakash, O. (2009). Effect of *Gluconacetobacter diazotrophicus* and *Trichoderma viride* on soil health, yield and N-economy of sugarcane cultivation under subtropical climatic conditions of India. *European Journal of Agronomy,* 30(4):296–303. doi: 10.1016/j.eja.2009.01.002.

Yedidia, I., Srivastva, A. K., Kapulnik, Y. and Chet, I. (2001). Effect of *Trichoderma harzianum* on microelement concentrations and increased growth of cucumber plants. *Plant and Soil,* 235:235–242. doi: 10.1023/A:1011990013955.

Zhang, S. W., Gan, Y. T. and Xu, B. L. (2015). Biocontrol potential of a native species of *Trichoderma longibrachiatum* against *Meloidogyne incognita. Applied Soil Ecology,* 94:21–29. doi: 10.1016/j.apsoil.2015.04.010.

Zhao, K., Penttinen, P., Zhang, X., Ao, X., Liu, M., Yu, X. and Chen, Q. (2014). Maize rhizosphere in Sichuan, China, hosts plant growth promoting *Burkholderia cepacia* with phosphate solubilizing and antifungal abilities. *Microbiological Research,* 169(1):76–82. doi: 10.1016/j.micres.2013.07.003.

Zhu, H., Ma, Y., Guo, Q., Xu, B. (2020). Biological weed control using *Trichoderma polysporum* strain HZ-31. *Crop Protection,* (134), doi: https://doi.org/10.1016/j.cropro.2020.105161.

Index

S

T

W